BADGES & UNIFORMS
OF THE ROYAL
AIR FORCE

BADGES & UNIFORMS
OF THE ROYAL
AIR FORCE

MALCOLM C. HOBART

Pen & Sword
AVIATION

In Memory of

JOAN
LUCY & CHARLIE

First published in Great Britain in 2000 by Leo Cooper

Reprinted in this format in 2012 by
PEN & SWORD AVIATION
An imprint of
Pen & Sword Books Ltd
47 Church Street
Barnsley
South Yorkshire
S70 2AS

Copyright © Malcolm C. Hobart, 2000, 2012

ISBN 978 1 84884 894 8

A CIP catalogue record for this book is
available from the British Library

Printed and bound in India by
Replika Press Pvt. Ltd.

Pen & Sword Books Ltd incorporates the Imprints of Pen & Sword Aviation,
Pen & Sword Family History, Pen & Sword Maritime, Pen & Sword Military,
Pen & Sword Discovery, Pen & Sword Politics, Pen & Sword Atlas,
Pen & Sword Archaeology, Wharncliffe Local History, Wharncliffe True Crime,
Wharncliffe Transport, Pen & Sword Select, Pen & Sword Military Classics,
Leo Cooper, The Praetorian Press, Claymore Press, Remember When,
Seaforth Publishing and Frontline Publishing

For a complete list of Pen & Sword titles please contact
PEN & SWORD BOOKS LIMITED
47 Church Street, Barnsley, South Yorkshire, S70 2AS, England
E-mail: enquiries@pen-and-sword.co.uk
Website: www.pen-and-sword.co.uk

Contents

INTRODUCTION

Military badge collecting has been a hobby for many, including the author, for some time. Collections of brass, anodised and cloth badges became popular following the Second World War and many military fairs are held throughout the country.

While serving in the Air Training Corps I became interested in RAF badges. I was able to gain information at RAF stations, but while researching the history I was unable to find a dedicated book on the subject.

Privatisation is gradually replacing regular Air Force personnel in today's Royal Air Force, with the result that uniformed staff of certain trades will soon no longer wear their trade badges. These will become obsolete and disappear into history.

In this volume, I have set out, to the best of my knowledge, a definitive guide to badges worn by the Royal Flying Corps, Royal Air Force and auxiliary units. I hope the book will assist both new or experienced collectors and the aviation enthusiast alike.

MALCOLM HOBART
APRIL 2000

CHAPTER ONE

THE ROYAL FLYING CORPS 1912-1918

Uniform Notes

When The Royal Flying Corps emerged from the Air Battalion of the Royal Engineers, on the 13th May 1912, the uniforms worn were either standard Army or Naval patterns. In the spring of 1913, permission was granted by King George V to dress the new military Flying Corps in a distinctive uniform with new badges. This decision was not to be made official however until the publication of Army Order 378 of November 1913.

The new service uniform for officers consisted of a Khaki folding cloth cap, similar to the later adopted RAF (FS) forage cap, being boat shaped with two brass buttons at the front and the RFC badge on the forward right side,(as viewed). The tunic was Khaki and double breasted with a high stand or fall collar and cut in a similar style to a lancer's jacket, with hidden buttons, and worn with Bedford cord breeches. Puttees with brown ankle boots were worn, although some officers continued to wear their Army service boots. The Sam Browne belt was also retained. Potential flyers were being seconded from Army regiments and many of them retained their uniforms, with the addition of RFC cap and collar badges. Warrant Officers wore officers' uniforms with the Sam Browne belt, but with airmen's cap and uniform badges. Other Ranks' uniform closely resembled the officers' uniform, but with jacket and breeches in serge material. Puttees and black ankle boots were worn.Due to the dirty nature of an air mechanic's job blue "boiler suit" overalls were found to be most practical for working dress.

Tropical dress consisted of a single breasted jacket without belt or skirt pockets in Khaki drill. Worn with long shorts which could be turned up and secured by a single button on the outer side of the leg. On the lower leg Khaki puttees with service boots were the main footwear. Tropical helmets or more rarely side hats were worn.

A full dress uniform was devised and authorised at the same time, but few seem to have been manufactured or worn.

For officers the uniform consisted of a dark blue cap with leather peak and scarlet band. The jacket was single breasted with eight buttons and scarlet high standing collar, edged in dark blue braid. The rear skirts were adorned with a button and three pieces of Russian braid in dark blue, on each side. Pointed scarlet cuffs peaked at six inches from the sleeve edge and gilt twisted cord epaulette were attached to the shoulders. A dark blue three inch Petersham belt was worn, which was fastened by three gilt wire toggles.

The Other Ranks Uniform was essentially the same but with detail variations. The main ones being seven buttons on the jacket front and shoulder boards bearing brass RFC titles.Rank badges were in gold lace for chevrons with embroidered gold badges.

Trousers for both officers and men were dark blue with a two inch (50mm) scarlet stripe down the outer seams and worn with black shoes or boots.

Women's Legion

The formation of the Women's Legion in 1915, which allowed the Army to recruit women, for non combat military duties, allowed many of the ladies to serve in the Royal Flying Corps. Their duties ranged from clerical to motorcycle dispatch riders. These women wore a loose fitting light Khaki coat, which had skirt pockets and was fastened by two RFC buttons and a cloth waist belt. The hat worn followed the fashion of the time and was "pudding Basin" in shape,with narrow brim. For normal duties skirts were long and full, but when more freedom was required, such as motorcycle riding, Light Khaki cord breeches were worn. Shoulder titles appeared on the upper arm and were the same as airmen's. Buttons for the above uniform were brass, with raised RFC letters and Kings Crown above.

Royal Flying Corps Badges

OFFICERS SERVICE DRESS

Cap Badge

The Royal Flying Corp cap badge closely followed the design of the Royal Engineers, from which it was formed. It consists of a laurel wreath broken at the top by a Kings Crown. The centre containing a monogram of the letters RFC, the monogram is pierced. For Officers the badge is made of bronze and is 1.5 inches (40mm) in size.

Collar Badges

These are replicas of the cap badge and about two thirds in size. They were worn in pairs on the step of the collar of the 1916 pattern jacket and sometimes on the stand-up collar of the 1908 jacket.

Shoulder Titles

Worn on the shoulder epaulettes of the jacket. These were brass capital letters 'RFC' connected at the top and bottom. The title being slightly curved in form.

Rank Badges

Being derived from the military wing of the Air Battalion of the Royal Engineers, the RFC followed the same rank structure, showing pips, crowns etc, on the shoulder straps of the RFC 'maternity' jacket and the cuffs of the 1908 Army tunic.

Flying Badges

Army Order 40 of February 1913 authorised what was to become the most sought after and recognisable of all flying badges, the pilots' wings. The design has changed little to the present day. Outstretched wings 4 inches(100mm) from tip to tip with ten to twelve feathers in

cream silk. These support a bronze embroidered laurel wreath encircling a RFC monogram, which is surmounted by a King's Crown in cream silk. These are embroidered on a shaped black patch. A gilt version was authorised by AO 40 and was attached to the Full Dress uniform by a broach pin.

The wings are supposed to represent swifts' wings.

Observers' Brevet

With the observers in the "back seat" sharing the same risks, a lot of these flyers felt deprived with the introduction of a badge for pilots only. This, no doubt, filtered back to the Army Council, but it was not until September 1915 that Army Order 327 announced a badge for qualified officers. This half wing or brevet showed an "O" attached to a wing extending to the right, as viewed, in cream silk. These being embroidered on a black, shaped background.

NCO's & OTHER RANKS SERVICE DRESS

Cap Badge
The cap badge was the same as officers, only in brass. A laurel wreath broken at the top by a KC. In the centre the RFC monogram. The badge being 1.5 inches (38mm) in diameter. Sometime after 1915 an unpierced solid version was issued as an economy measure.

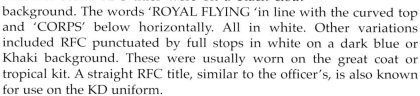

Shoulder Flash
Shoulder titles followed the standard Army pattern, being curved at the top to follow the shoulder seam. RFC titles were on a black cloth background. The words 'ROYAL FLYING 'in line with the curved top and 'CORPS' below horizontally. All in white. Other variations included RFC punctuated by full stops in white on a dark blue or Khaki background. These were usually worn on the great coat or tropical kit. A straight RFC title, similar to the officer's, is also known for use on the KD uniform.

Rank Badges
As with the officers, NCO rank followed the Army style, although the addition of propellers and stars started the move towards the flying services' new identity. Warrant Officers continued with their normal rank badges and chevrons were worn by sergeants and corporals. A new rank was created of Flight Sergeant. Lance corporals were replaced by Air Mechanic 1st Class. Rank badges were worn on both

upper sleeves except Warrant officers, these being on the lower sleeve 8 inches (200mm) from the cuff edge.

WARRANT OFFICERS 1ST CLASS OR RSM.
Royal coat of arms in Khaki worsted.

WARRANT OFFICERS 2ND CLASS
2 inch (50mm) crown, pale Khaki worsted on a Khaki shaped patch.

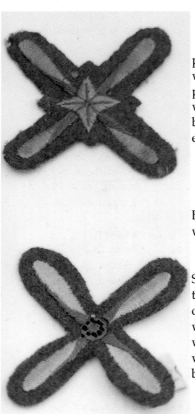

FLIGHT SERGEANT
A four blade propeller on which a four pointed star was superimposed on the boss, with above, a Kings Crown. Crown and propeller in white, cream or buff coloured worsted on a shaped Khaki patch. This badge was worn above three chevrons also either white, cream or buff.

QUARTERMASTER SERGEANT
Badge as above, but worn with four inverted white, cream or buff chevrons.

SERGEANT
Sergeants were recognised, as always, by three chevrons. These were in white or buff on a Khaki background. In addition they wore above these a shaped patch on which was embroidered a four bladed propeller in white or buff. The centre having a round boss.

CORPORAL
Corporals retained the two chevrons in white, cream or buff.

AIR MECHANIC 1ST CLASS
A twin bladed propeller in white, cream or buff silk, with a circular central boss, embroidered on a Khaki patch. Worn after October 1916, Army Order 322.

Flying Badges

At this time, apart from the pilot, the only additional aircrew were Observers, increasingly Other Ranks were being used for this duty. Following the introduction of the 'Observers Brevet' for officers, Army Order 404 of November 1915, authorised the same badge to be worn by qualified WO's, NCO's and men.

The badge is as described for officers.

Trades

With the advent of the early form of wireless, personnel were needed to be trained in servicing and using this new apparatus. A brass badge appeared which showed an 'O' with three lightning flashes emitting from each side. This would appear to have been an unofficial badge. The author has been unable to find any photographs or Army Orders to verify the badge's authenticity.

While personnel were transferring to the RFC from other branches of the Army, many men retained qualification badges from their previous regiments. Photographs of this period show suitable qualifications eg signaller, to unsuitable eg skill-at-arms with a lance, being worn.

RFC FULL DRESS UNIFORM

Officers

As previously mentioned, a Full Dress uniform had been authorised by Army Order 378 in November 1913. By the time the order was promulgated war was threatening and it would appear that very few Full Dress uniforms were made or worn. Photographs showing this uniform are very rare.

Cap Badge
A laurel wreath,pierced at the top by a K/C, encircling the RFC monogram. The same as the Service Dress badge, only in gilt metal.
Collar Badge
As cap badge, only two thirds the size, in gilt metal.

Flying Badge
The only flying badge worn on Full Dress were Pilots' Wings.These being in gilded silver and attached to the uniform by a brooch pin.

NCO's & Other Ranks
The uniform was identical to the officers with detail variations. Gilt metal RFC initials, on plain dark blue shoulder straps, replaced the gilt cord epaulettes.

Cap Badge
A laurel wreath pierced at the top by a King's Crown, encircling the RFC monogram, in brass.

Collar Badge
In brass as above, only 2/3rds the size.

Shoulder Titles
These were in gilt metal, RFC initials with connecting bars top and bottom.

Ranks
Badges of rank were worn on the right sleeve only. Warrant Officers 1st & 2nd class and Quartermaster Sergeants displayed their rank on the lower sleeve. Flight Sergeants and below showed their rank on the upper sleeve. Chevrons were made of gold lace with badges in gilt thread.

Rank Badges

Warrant Officer 1: Royal Coat of Arms in gilt thread.

Warrant Officer 2: Large gilt embroidered Crown

Quartermaster Sergeant: Four bladed propeller with star boss, above four inverted chevrons.

Flight Sergeant: Four bladed propeller with star boss, with Crown above ,worn above three chevrons.

Sergeant: Three chevrons and a four bladed propeller with round boss.

Corporal: Two chevrons.

ROYAL WARRANTS

20
Royal
Flying
Corps

Royal Flying Corps (Military Wing)

No. 130
1912

GEORGE R. I.

WHEREAS WE have approved of the establishment of an aerial service for naval and military purposes under the designation of the Royal Flying Corps;

AND WHEREAS it is necessary to form a Military Wing of the Royal Flying Corps to which officers and men of Our Land Forces can be appointed; OUR WILL AND PLEASURE is that the Royal Flying Corps (Military Wing) shall be decreed to be a corps for the purpose of the Army Act.

Given at Our Court at St James's this 13th day of April, 1912, in the Second year of Our Reign.

By His Majesty's Command,

HALDANE OF CLOAN

CHAPTER TWO

THE ROYAL AIR FORCE

April 1918 - September 1919

During the middle of 1917, the German Zeppelin raids over Britain had brought the British Government to rethink the defence of the British lsles. At this time the defence of the Home waters and coast was the responsibility of the Royal Naval Air Service while air cover on the continent was supplied by the Royal Flying Corps and some units of the RNAS. Aircraft production was also increasing and it was clear that improved organisation and logistics of the flying service was required, to make better use of resources.

To this end, the British Government of the day brought in the South African statesman General Jan Smuts to study the problem and put forward his recommendations for improvement. He produced a report on the 19th July 1917, recommending among other things, the creation of a unified Air Command of Britain. A second report dated 17th August went further and suggested a complete reorganisation of the entire air arm. This suggestion met with mixed response from both the Army and Navy. Both services wanting to retain their own air command. The timing was also in question, being at the height of hostilities. However the Government reacted quickly and the 'Air Force Act' was put before Parliament and received Royal Assent. An Air Council was created on the 2nd January 1918, which merged the Military and Naval air arms together and thus formed the Royal Air Force on the 1st April 1918. Orders for the new Royal Air Force now came from the Air Ministry and up to 1930 were prefixed AMWO (Air Ministry Weekly Orders) and from 1930 AMOA (Air Ministry Order Administrative).

ROYAL AIR FORCE SERVICE DRESS
UNIFORM NOTES

1st April 1918 -15th September 1919

Instructions for the new uniform and badges were set out in AFM 2, and described the use of two uniform colours for Officers and Other Ranks. Continuation of the use of Khaki and a new pale blue/grey. A Khaki uniform was to be issued to Airman for the duration of the war or until stocks were exhausted. As the Officers bought their own uniform, a wearing out period was allowed. It can be realised therefore that the transformation to the new uniform was slow, especially overseas.

Khaki Uniform (Officers) (K)

The Khaki uniform consisted of a single breasted,four buttoned jacket, with two breast pockets and patch skirt pockets. The pockets were flapped and fastened with a single button. There were no shoulder straps and the cloth belt was attached at the rear. Buttons were gilt metal,with a raised eagle and bordered by a rope edging.The jacket was open necked, exposing a Khaki shirt and black tie. Matching trousers or breeches with puttees were worn with brown shoes or boots. The uniform was finished with a Khaki cap which had a black mohair band and leather chin strap.

NCO's & Other Ranks

The Other Ranks uniform also had breast and skirt pockets, although the skirt pockets were internal and the flaps were not buttoned. It was made of serge material with the exception of Warrant Officer I, where the uniform was of officer cut and quality and worn with a shirt and tie. Other NCO's and men fastened the tunic to the neck. Possibly because of war time restrictions, the original gilded metal buttons were replaced by brown leather ones. Metal buttons did not return until May 1920. Brown boots with puttees were worn for best dress, but trousers and shoes were allowed for normal duties on camp.

Although originally intended to be worn for the duration of the Great War, with the intention of replacing Khaki by the new pale blue uniform thereafter, the Khaki uniform was worn by Other Ranks until

July 1924, when AMWO 484 was published,although mainly as working dress.This order stated that all Khaki uniform was to be returned to stores and replaced with the familiar blue/grey uniform.

The blue/grey uniform had previously been sanctioned by AMWO 1150 of October 1919, to be worn on parade and walking out use, when available.

Pale Blue/grey Uniform

Various rumours circulated at the time, as to why this shade of material was to be introduced. These ranged from stocks abandoned from an Imperial Russian Cavalry order to it being a favourite colour of a well known musical actress, who had friends in high places.

As already stated Khaki uniform was to be worn until the end of the 1914-18 War, but officers were allowed to wear the pale blue/grey uniform for mess dress. The design of the jacket was the same, only with gilt metal buttons bearing an eagle only and still worn with puttees or trousers. Shirts were silver grey or white according to AMWO at the time. The cap was the same, but in pale blue/grey. Footwear now changed to black.

The transition from pale blue/grey to blue grey uniform was promulgated on the 15th September 1919 by Air Ministry Order 1049.

Badges and rank, although the same were made of different materials and colour. These will be explained in the badge section.

KD Tropical

Tropical Kit was authorised in July 1918 and consisted of a tunic in Khaki Drill with detachable shoulder straps and worn with slacks for working dress, with breeches and field boots for parade uniform. Shorts were officially allowed in the spring of 1921, (AMO 353) depending on local conditions.

The Womens Royal Air Force

Although the Women's Royal Air Force came into being at the same time as the RAF, 1st April 1918, the provision and organisation was more chaotic than it's male counterpart. Uniforms that were originally issued were either as used by women in the RFC or the Women's Army Auxiliary Corps.

As with the men Khaki prevailed in the original uniforms, which consisted of a loose fitting open necked jacket , mid calf full skirt, black shoes and Khaki stockings. A soft crowned cloth peaked cap finished the uniform. The jacket was single breasted with four metal buttons and had an attached cloth belt, with a single button fastening. Two pockets were provided on the jacket skirts with buttoned flaps. A white shirt with black tie was worn.

During 1919 some issues of the new blue uniform were made available, as the WRAF slowly came fully into being. A long double breasted jacket secured by four leather buttons on the left side and fastened to the neck. The attached belt was secured by a single button, skirts were full and worn with the hem 12 inches from the ground. Shoes were black. The cap was of the same design as the Khaki uniform, soft crown with pleating at the sides and cloth peak.

Cap Badge:
A domed, black circular patch with K/C above. Centrally the RAF eagle surrounded by a rope border embroidered in white.

Officers' badges followed their male counterparts, having cuff lace with the gilt eagle above. OR's wore shoulder titles with white 'W.R.A.F.' letters on a black curved, shaped patch, this was then replaced by the shoulder eagle in white on a black, shaped patch. In some photographs both badges are worn.

Before the wearing of chevrons, Chief Section Leaders and Section Leaders were denoted by the wearing of badges below the eagle.

Chief Section Leader:
A crowned eagle with below twin laurel sprigs in off white, on a black shaped patch.

Section Leader:
As above, without the crown.

A large white capital 'I' was worn by some personnel to denote that they worked only locally to home and were therefore in the 'Immobile Branch'.

By the 1st April 1920, the Women's Royal Air Force had ceased to exist and apart from medical and clerical services, women were not

recalled for service until the Women's Auxiliary Air Force was formed on the 28th June 1939.

ROYAL AIR FORCE SERVICE UNIFORM BADGES

1918 - 1919

With the creation of this new service a new emblem was needed. The obvious choice for air? A Bird. During the passage of time, various sources have suggested the Royal Air Force (bird) is either an eagle or an albatross,which?

Air Force Memorandum No 2 of 1918 states that the badge worn on the officers' cap will be as follows 'The badge is entwined laurel leaves, above which is a reduced facsimile of the metal bird at present worn on the sleeve by the R.N.A.S. (Royal Navel Air Service) ,the whole being surmounted by a Crown'.

The eagle was first adopted by the Royal Naval Air Service in July 1912, following the directive of Winston Churchill, that Naval airmen were to wear a distinguishing eagle badge on both cuffs of their uniform jacket. The original artist's design failed to impress and it was the wife of Admiral Murray Sueter, the executive officer charged with

the creation of the badge, who suggested copying a brooch she owned of a French Imperial eagle design. Admiral Prince Louis of Battenburg and Mr Churchill agreed that it was of a more suitable design and it was adopted.

As already stated, during the transitional period from the Royal Flying Corps to the Royal Air Force, two colours were in use for uniform. These were light blue/grey and Khaki and such colour variations occurred in some badges. Where this occurs in the following text a code will be used as follows.

Light blue/grey   (LB)   Khaki   (K)

OFFICERS' BADGES

Cap Badge
AMWO 617 1918
Air Officers: Gold wire, embroidered laurel leaf wreath bordering a black padded patch and broken at the top by a K/C in full colour. Above the Crown a Crowned lion in gold thread. In the centre a gilt metal eagle with wings superimposed on the laurel wreath.

Officers:	Three patterns were used as follows.
1st pattern	Centrally a gilt eagle supported by four gold wire laurel leafs below. Above the eagle a gold embroidered crown with red cushion. All on a black cloth backing.
2nd pattern	A one piece, all metal badge containing all the elements as above. (Later worn by Warrant Officers only).
3rd pattern	As above, only in three sections and fixed to a black cloth backing.

Second Lieutenants and Lieutenants wore single, gilt metal bars on either side of the cap badge, with Captains wearing two bars each side.

In addition, Majors to Colonels showed a single row of oak leaves on the cap peak and two rows of the same were worn by General Officers.

Rank Insignia

Now that the Royal Air Force was completely independent of the other Services, it was desirable to emphasise this fact. Air Force Memorandum 2 1918 set out the new distinctions, although the position of the rank on the jacket cuff was to be retained, an adaptation of the Naval cuff ring was to be used in three sizes of braid. These being 1/4 inch, 1/2 inch and 2 inches. The different combinations denoted different ranks as follows.

General :   One 2 inch with three 1/2 inch above.

Lieutenant General: One 2 inch with two 1/2 inch above.

Major General: One 2 inch with one 1/2 inch above.

Brigadier General: One 2 inch.

Colonel: Four 1/2 inch.

Lieutenant Colonel: Three 1/2 inch.

Major: Two 1/2 inch with one 1/4 inch between.

Captain: Two 1/2 inch.

Lieutenant: One 1/2 inch.

Second Lieutenant: Crowned eagles only.

These eagles were replaced in August 1918 by a 1/4 inch cuff lace (AMWO 617). The colouring differed on the two uniforms. (LB) Gold braid. (K) Blue grey central stripe on a Khaki Braid.

Above the braid, a crowned metal eagle was worn. These were

supplied in pairs, the eagle always 'flying' to the rear. The Crown and eagle were separate, being spaced by a backing plate.

NCO's & OTHER RANKS

Cap Badge

Warrant Officer I: As described for Officers, pattern (2)
(Originally as Officers' pattern (1)

Warrant Officer II: A circular black domed velvet cushion backing with a raised attached crowned shaped patch. Centrally an angular eagle within a rope border, with K/C above. Detail in gold wire.

Airmen:
As above, but embroidered in red worsted

AMWO 783/19 announced a new cap badge. This being the familiar laurel wreath pieced at the top by the K/C, with the RAF monogram in the centre. It was worn by ranks below Warrant Officer I.

Shoulder Eagle: The shoulder title was replaced by an eagle badge, issued in pairs and worn below the shoulder seam, facing to the rear.
(K) Khaki oblong patch with an eagle in red silk.

(LB) Black or dark blue rectangular patch with eagle in light blue silk.

Rank Badges (K)

Warrant Officer I :	Gilded metal eagle and Crown. (Worn on both upper sleeves).
Warrant Officer II:	Large Khaki Worsted or gilt Crown. (Worn on lower sleeves).
Flight Sergeant:	Three Khaki Worsted chevrons with large Khaki Worsted or gilt Crown above. (Worn on upper sleeves).
Sergeant:	Three Khaki Worsted chevrons. (Worn on upper sleeve)
Corporal:	Two Khaki Worsted chevrons. (Worn on upper sleeves).
Air Mechanic 1st class :	Red silk twin bladed propeller on a Khaki rectangular or shaped patch. (Worn on the upper arms).
Rank badges (LB):	As promulgated in AMWO 728/18 of 1918.
Warrant Officer I:	Royal Coat of Arms in L/blue silk with red crown cushion. (Worn on the lower sleeve.)

Other Ranks remained the same only in light blue on dark blue patches.

CHAPTER THREE

THE ROYAL AIR FORCE   1920 – PRESENT

Uniform Notes

In the later part of 1919, attention was concentrated in improving the identity of the Royal Air Force. To this end in August 1919 AMWO 973 announced a change in naming of the commissioned officers ranks, although the structure was to remain the same.This was followed in September,by AMWO 1049 standardising the uniform colour as blue/grey. Originally the highest rank was to be 'Marshal of the Air', however within a week, at the request of King George V, the rank was changed to 'Marshal of the Royal Air Force'. The new rank names were as follows:

	Marshal of the Royal Air Force	(1)
General	**Air Chief Marshal**	(2)
Lieutenant General	**Air Marshal.**	(3)
Major General	**Air Vice Marshal**	(4)
Brigadier General	**Air Commodore**	(5)
Colonel	**Group Captain**	(6)
Lieutenant Colonel	**Wing Commander**	(7)
Major	**Squadron Leader**	(8)
Captain	**Flight Lieutenant**	(9)
Lieutenant	**Flying Officer**	(10)
Second Lieutenant	**Pilot Officer**	(10a)

Rank lace also changed to black with a bright, pale blue central stripe. The combinations of size and position remained as before.

Headdress

During 1920 a new cap was introduced with a fabric peak for Officers. Air Officers therefore discontinued wearing oak leaves on the peak until 1932, when on the 22nd November AMWO 355 re-introduced a leather peak for Air Officers and Group Captains. The leather peak allowed Air Officers to show two rows and Group Captains one row of gold oak leaves.

A new form of headgear was brought into service in 1936: the Field

RANK DISTINCTIONS IN THE
ROYAL AIR FORCE

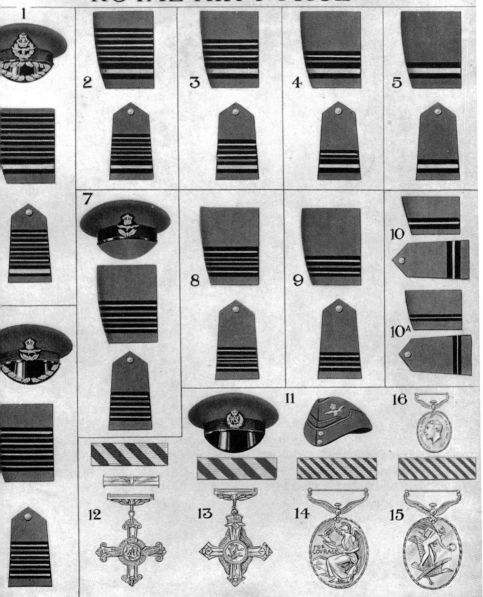

Service Cap. A folding cloth boat-shaped cap, with two gilt metal buttons at the front, and the ear flaps along the side of the cap, (AMO A93). In 1940 (AM0 A 142/40) brought into use a special FS cap for Air Officers, which had pale blue piped edges.

In December 1939, all Non-Commissioned ranks were instructed to wear the Field Service cap only, as the SD cap was withdrawn. There were exceptions, as MT drivers, Police and Apprentices still retained the SD cap.

The RAF Regiment were the first to wear the familiar beret, in 1943. This form of headgear became increasingly used and was adopted by other trades during the war, to become the standard wear afterwards.

Service Dress

The Service Dress jacket has remained in use since 1918 and is still worn as the No 1 uniform today. Some modifications have been made to bring the jacket up to date.

In 1946 a committee had met to review the Royal Air Force Post War Service Dress and one of the decisions made was to alter the design of the Officers' Home Service jacket. The new 1947 design had two vents at the rear and was fastened by three gilt metal buttons at the front. An unpopular feature of the new design was the removal of the side skirt pockets. The new tunic was not compulsory, and the wearing of the old style jacket was allowed until the need for replacement. The unpopularity of the jacket prompted a return to the original style in 1951.

The Officers' jacket was single breasted with four gilt metal buttons and a detachable fabric belt. The modern jacket has three gilt anodised aluminium buttons with a forth flat black plastic button hidden by the belt buckle. All pockets have flaps,which are secured by a gilt metal button. The skirt pockets are of the external patch type. Officer and Warrant Officer uniforms are made of barathea material and worn with a wedgewood blue shirt, black tie and black shoes. Brown leather gloves complete the No 1 uniform

In 1936 a similar jacket was authorised for Other Ranks,being made of blue/grey serge with four gilt metal buttons, but no belt. A B/G webbing belt was worn on parade. Only the breast pockets have buttoned flaps with the lower pockets, which are of the internal type, being buttonless. Light blue shirts, black ties and black shoes or boots were worn.

War Service Dress

Following the successful adoption of a blue/grey version of the Army Battle Dress for Air Crew in December 1940 (AMO A909), a decision was made to use this garment as standard uniform for all ranks. Known as War Service Dress its use started in 1943 and was authorised by AMO A 1062.

War Service Dress continued to be worn after the Second World War as No 2 Home Service Dress, following significant alterations in 1948 and was to be worn until 1973. A waist length jacket, fastened at the front by an extended waist band and buckle, with black plastic buttons covered by a fly front and two pleated breast, flapped, pockets. Officers' rank appeared on the shoulder straps.

Heavy Duty Dress

Heavy Duty Dress was the RAF's terminology for the Army Khaki Battle Dress and was worn mainly by the RAF Regiment. Other personnel to wear HDD were Glider aircrews, Airfield Construction crews, Combined Operations and aircrews of the 2nd Tactical Air Force. Officers wore the jacket open at the neck revealing a blue shirt and black tie. Airmen were ordered to button the jacket to the neck (AMO A 251/42). The RAF Regiment was allowed to wear Khaki shirts from January 1944.

War Service Dress was withdrawn from service in the early 1970's, although Heavy Duty Dress disappeared soon after the War.

Modern Service Dress

Uniforms in today's Royal Air Force, as with other services, follow strict uniform code. To avoid confusion when published in Standing Orders, modes of uniform dress are numbered and follow the code shown below.

No 1 Dress: BG jacket and trousers, light blue shirt with black tie, peaked cap, black shoes and brown leather gloves for officers and Warrant Officers.

2 Dress :   BG zipped jacket, trousers, light blue
(obsolete)   shirt with black tie, beret or cap.

No 2A Dress :   Woollen crew necked jumper (now V necked

BG trousers, black tie and shoes.
Cap or Field Service Cap (Officers & Warrant
Officers). Beret (Other Ranks)

No 2B Dress: Long sleeve light blue shirt with black tie, BG trousers
and black shoes. Short sleeved light blue shirt,
unbuttoned at the neck, BG trousers and black shoes.
RAF Stable Belts may be worn with above 2B Dress.

No 3 Dress: Nato Combat Dress

    Rank on No1 Dress and No2 Dress is shown on the sleeve. On No2A
and 2B Dress rank is worn on woven shoulder slides. No2 Dress (1972
pattern) jacket had a high necked folded collar with a central blue
nylon zip front. Breast zipped pockets and shoulder straps.

    In 1996 a new wind-cheater style jacket was issued to replace the
1972 pattern, which was withdrawn in 1992. In dark blue
cotton material the jacket retains epaulettes for rank slides and has a
zipper fastening and two side pockets. Velcro tabs allow a lining to be
fitted.

    No2A Dress woollen pullovers are ribbed with blue grey fabric
shoulder and elbow patches and cloth epaulettes for rank slides. The V
necked, later version has a patch on the left upper sleeve to hold two
pens.

    A new unisex pullover was issued in 1996 which is V necked, with-
out ribbing and retains the shoulder and elbow patches,as well as the
epaulettes for rank slides. The pen pocket now has a retaining, velcro
fastened flap.

    Dark blue shirts are sometimes used for working dress.

Greatcoats

    The greatcoat had been inherited from the Army in 1920 and were
now in blue grey. Officers and WO's greatcoats had a fabric belt with
brass buckle and two rows of five buttons. Three smaller Brass buttons
were worn on the sleeve cuff. Shoulder boards were attached for officers and showed the rank braid. Other Ranks greatcoats had shoulder
straps and no belts, webbing belts being worn on parades, and had two
rows of four buttons.

The Officers' coat had originally been made of a fleece cloth, but was changed to Melton in 1942. Airmen had to contend with the ordinary serge material.

Modern

Raincoats have superseded greatcoats for normal wear in todays Royal Air Force. Greatcoats are still worn on ceremonial occasions, by Officers of Air Rank, Bands and members of the RAF Regiment.

Officers' and OR's raincoats are calf length and have three large black plastic buttons and a cloth belt fastened with a black plastic buckle. There are shoulder straps to show rank. The previous 1972 pattern raincoat was without a belt for all ranks. This design did not have shoulder straps and Other Ranks' rank was sewn on to the sleeve. Officers displayed their rank by the use of enamel badges, pinned to the lapels.

Tropical Dress

Involvement in India, the Middle East and Far East had been undertaken since the RAF's early days. Tunics were of similar design to the Home Service Dress, but in Khaki Drill cotton. Trousers or shorts were worn with fawn socks and black shoes. Shirts were also in Khaki Drill colour and when a tie was worn it was black. The Bush shirt, which had two breast pockets, was often worn without the jacket and in time developed pockets in the skirts. Normal Home Service headgear was worn,with tropical helmets issued until 1942. Officers' Rank was shown on the jacket/shirt shoulder straps with Warrant Officers wearing a gilt badge on a brown leather strap, on the right wrist. NCOs continued to wear chevrons on the upper sleeve, but these were of white tape.

Modern

No1 Dress (Warm Weather Areas) is made of a Khaki Drill material. The tunic jacket for Officers and OR's is single breasted with four anodized aluminium (staybright) buttons and a fabric belt. Officers' pockets are buttoned as the No1 Home Service Dress, with shoulder straps to display rank slides. The OR's jacket shoulders are plain. The uniform is completed with Khaki Drill trousers and shirt with black tie and shoes.

Working Dress can be a long sleeve shirt, tie and trousers with black shoes (No 7A Dress) or Short sleeved shirt, shorts, long socks and desert boots (No 7B Dress).

WOMENS AUXILIARY AIR FORCE

Following the successful participation of women during the Great War, the Air Ministry declared the formation of the Women's Auxiliary Air Force on the 28th June 1939. Originally WAAF members were to release RAF personnel from administration and domestic duties but as the Second World War progressed, women were to be found in many trades.

The uniform consisted of a jacket similar to the men's, but tailored to the female shape. It had four brass buttons and a fabric belt. All pockets were flapped and fastened with brass buttons until, AMO N 1108 of October 1943 discontinued the buttons on the lower pockets. Skirts were plain and straight with the lower hem terminating 14 to 16 inches from the ground. The uniform was made of barathea-like material and worn with a light blue shirt for Officers and blue grey for OR's. Black ties and black lace up flat shoes were worn.

The crown of the Officers' cap was made of three sections of fabric with attached cloth peak, all in B/G. Other Ranks' caps were constructed in one piece drawn into a black hat band with a patent leather peak. Both designs had a false, fabric rear peak.

War Service Dress (termed 'Suits Working Serge' in the WAAF), slacks and overalls were all worn during the war as duties required. Greatcoats and raincoats were also issued.

Women's Auxiliary Air Force Officers wore rank braid the same as their male counterparts. The titles were however different until 1968, when the more familiar rank titles were introduced. The comparison chart below sets out the changes and the RAF equivalents.

WAAF	RAF
Air Commandant	Air Commodore
Group Officer	Group Captain
Wing Officer	Wing Commander
Squadron Officer	Squadron Leader
Flight Officer	Flight Lieutenant

Section Officer   Flying Officer
Assistant Section Officer   Pilot Officer

At a latter stage in the WAAF development, equivalent ranks of the Royal Air Force's Air Vice Marshal and Air Marshal were added,to be known as Commandant in Chief and Air Chief Commandant respectively.

In 1942 NCO's and OR's began to wear the same rank and trade-badges as RAF personnel. Before January 1942, the Highest Non Commissioned Rank was known as Under Officer and a Crown badge was worn. After this date the rank title was changed to Warrant Officer as used by the RAF and cap and rank badges were to be the same (AMO A104).

WOMENS ROYAL AIR FORCE

On the 1st February 1948, the Women's Auxiliary Air Force was renamed the Women's Royal Air Force, following the Royal assent of the Army and Air Force (Women's Services Act).

The headgear changed to a stiff, high fronted cap with a black leather peak and chin strap during the 1950's. This was followed in the sixties by a softer cap similar to the RAF Officers' cap. Other Ranks hat became based on an 'Air Hostess' design, being of an oblong pill box shape. In the early 1990's a trilby style hat with the side and rear brim turned up and black mohair band was introduced and is still being worn. The beret continues to be worn with working uniform.

The Officers and OR's jackets are single breasted and waisted, with four gilt (staybright) buttons and no belt. Two flapped, pockets appear on the jacket skirt fastened with gilt buttons. Loose, straight skirts are worn with the hem below the knee. As with the men, V necked jerseys are worn. These are made of a softer wool and have wider ribs.

The female jersey does not have shoulder or cloth patches. Wedgewood blue shirts, black tie and black shoes complete the uniform. In 1997, the colour of tights changed from pewter to nearly black.

In 1997 a new 'unisex' V necked jersey was introduced for all members of the Royal Air Force. This is of a plain knit, with cloth shoulder and elbow patches. The left upper sleeve has a flapped pen pocket.

PRINCESS MARY'S ROYAL AIR FORCE NURSING SERVICE

The Princess Mary's Royal Nursing Service (PMRAFNS), can trace it's origins from the formation on the 1st June 1918, with it's Royal patronship being awarded in 1923. Although a dedicated Royal Air Force nursing service, the PM's were controlled by a medical board who's advice was followed by the Air Ministry. Personnel of the service were originally given appointments, as opposed to ranks, which followed the traditional structure of their profession.

Originally the nursing grades were shown by two types of maroon lace. Matrons and Sisters appointments were shown by a 1 1/2 inch lace with a 1/2 inch pale blue central stripe, with Senior Sisters and Sisters lace being 1 1/4 inch wide and having a 1/4 inch stripe. The lace was worn on the lower cuff in the same position as RAF Officers. On clothing that was without cuffs, maroon shoulder boards were worn with a 1/2 inch or 1/4 inch central tape. Staff Nurses wore dark blue shoulder boards.

With the introduction of the Defence (Women's Forces) regulations of 1941, in common with other Female Services, the PMRAFNS became integrated into the Armed Forces. In consequence, AMO A196 of March 1945, ordered that members of the PMRAFNS would wear RAF rank braid and badges, although they would retain their appointment titles to equal the Royal Air Force, as indicated below.

Matron in Chief	Air Commodore
Principle Matron	Wing Commander
Matron	Squadron Leader
Senior Sister	Flight Lieutenant
Sister	Flying Officer

The Staff Nurse rank had been abolished on the 1st April 1941 (AMO 505/41).

Entrance to the PMRAFNS had originally been by commissioned rank only, but in 1960 a decision was made to enlist student Nurses, to undertake a three year course to State Enrolled Nursing standard. The RAF Nurse Training School was disbanded in December 1995 with training now confined to the tri-Service Royal Hospital Haslar and Portsmouth University.

In a move to consolidate the RAF Medical Services the rank titles of the PMRAFNS were changed to follow the rank titles of the RAF and WRAF on the 1st April 1980.

Uniform

Summer Service Dress for PM's originally consisted of a Norfolk style jacket,with a three button fastening,button skirt pockets and a double buttoned attached cloth belt. This was worn with a matching skirt which reached to mid calf. The skirt was unusual in having broad pleats at front and rear. A choice of head-wear was available, being either a storm cap with false front and rear peaks or a black felt, four cornered hat, both bearing a black mohair band. This uniform was completed with black shoes and tie, white rounded collar shirt and gloves.

The Winter uniform was made of B/G Barathea and was in the form of a frock-coat, with long sleeves and a 12 half button front, which hid a press-stud fastening. A shoulder cape was often worn attached at the front of the low collar and decorated at the rear centre by a large rosette. Medical collar badges were shown on the front lower cape corners. A white veil,black laced shoes and black stockings were also worn with this form of dress.

On these long sleeved uniforms rank was shown by braid on the lower sleeve. When greatcoats were worn rank was indicated by coloured shoulder boards or later by B/G boards and braid. A three buttoned cloak was authorised in 1936, with a 2 inch rise or fall collar, which denoted matrons by a maroon collar and lining, with senior sisters and sisters showing a dark blue collar, the latter, in common with staff nurses, having powder blue linings.

A white cotton drill dress was worn for ward duties, which had elbow length sleeves, two shirt pockets and was buttoned at the front by 14 domed mother of pearl buttons. Once again, a veil was worn, but stockings were opaque and worn with canvas or buckskin laced shoes. Matrons continued to wear their B/G frock-coat.

In warm weather areas of service, the PM's dresses were white cotton with long sleeves and a 3 inch open collar. They were buttoned at the front with 14 pearl buttons, with the matron's dress being in trico-line and having 12 buttons. Head-wear could be either the veil, a white felt panama hat or a white solar helmet which peculiarly showed a blue coloured brim underside.

No1 Service Dress

Today Service Dress of the PMRAFNS is the same as other RAF personnel, although female officers and airwomen wear different headgear. Both in B/G material, the Officers' hat has a round crown top with a wide, raised brim which is lower at the front. The airwomen's hat also has a round crown but a smaller upturned brim.

No 2D Service Dress

This uniform is for ward use. Female staff wear a white short sleeved dress, buttoned at the front. Status is shown on shoulder rank slides, the officers wearing a short elbow length D/blue cape. A white veil headdress is worn by officers, while OR's wear the traditional white cap. White laced shoes complete the attire.

Male members of the PM's dress in a white 'maternity' style short sleeved shirt with epaulettes to show rank. This is worn with uniform trousers and black shoes. In Warm Weather areas white shorts and socks are worn. A standard SD cap is used for head-wear.

No 2E service Dress

Used by female personnel employed on aeromedical duties, this dress consists of a white open necked, short sleeved shirt, worn outside black slacks, with black shoes. D/blue tape decorates the lower sleeves and collar edging. Rank is shown on epaulette slides.

CHAPTER FOUR

CEREMONIAL, MESS DRESS AND BANDS

The Royal Air Force, being the Junior Service, did not have hundreds of years of tradition to draw from to design a uniform for ceremonial or important occasions. The reluctance to issue colourful, or indeed, many badges also hindered the grandeur drawn upon by its Sister Services.

Following the Great War, the Air Ministry was able to direct it's thoughts towards designing suitable garments for these occasions. As already stated, officers were originally allowed to wear the pale blue uniform for Mess Dress during the duration of the war, with the working jacket and trousers used for parade wear. These Orders of Dress continued until the early 1920's, when the following new clothing appeared.

Full Dress

In April 1920, AMWO 332, authorised a Full Dress uniform for Home Service wear. This consisted of a single breasted jacket in blue grey material, having a 1 1/2 inch stand up collar and a seven gilt metal button fastening. The rear skirts were also adorned at the top by two gilt buttons and further smaller buttons were used on the shoulder boards near the collar. The buttons were unique to the Full Dress and Mess Dress, being domed with a separate gilded eagle and Crown mounted on the apex.

Rank was indicated by lace on the lower sleeve. These were in the same arrangement as Service Dress, being combinations of 2, 1/2 and 1/4 inch gold braid. In addition, the collar had gold wire embroidered devices of oak leaves and acorns on both sides to indicate rank bands as follows.

Pilot Officer - Flight Lieutenant..Five oak leaves at the collar front.

Squadron Leader - Group Captain ....Oak leaves and acorns

extending from collar front for 3 1/2 inches around the collar.

Air Officers: As above, but extending all around the collar. The collar itself being bordered by gold lace.

Shoulder boards for all Ranks had a Crown and eagle in gold wire embroidered in the middle, with Air Officers having the additional embellishment of laurel leaves surrounding the shoulder board button.

Originally trousers were plain blue grey without turn-ups and worn with half Wellington boots. These were superseded in June 1928 by close fitting cavalry trousers with instep strap, made of stretch barathea-like material which formed over the uppers of the half Wellington boots when worn, (AMWO 538). Air Ministry Order A 311 of December 1934 introduced a new pattern overall trouser to be worn with Mess Dress and Full Dress. Of similar style as above but with the outer seams bearing a gold silk lace stripe with two bright pale blue runs within. Flight Lieutenants to Group Captains having 1 1/4 inch lace with 1/16 inch blue stripes and Air Ranking Officers 1 1/2 inch lace with 1/8 inch blue stripe. Pilot Officers and Flying Officers retained their plain trousers. When Full Dress was first announced in 1920, the headdress to be worn was a Service Cap of B/G Venentian cloth. By June 1921 a new form of headdress was ordered by AMWO 551. Supposedly to imitate the design of the original flying helmet, the helmet consisted of a back leather skull cap trimmed with black rabbit fur, ear flaps attached to the side. The front showed the leather of the cap with fur along the lower edge, with the back being fur covered except for two triangular areas of black leather. On the front of the helmet is attached a basket weave holder masked by the Royal Air Force badge in gilt and black. Into the holder, an ostrich feather plume in RAF blue was placed. To complete the adornment, a blue and gold cord hung across the helmet front, beneath the badge and a patent leather chin strap was attached below.

This attempt at a ceremonial helmet was never popular with the officers intended to wear it. Being expensive and seldom used, Junior officers preferred to hire the complete Full Dress uniform and were eventually allowed to wear Service Dress caps on Full Dress occasions.

A straight sword completed Full Dress, hung from a plain woven blue belt, with a gilt round tongue and slot buckle. The buckle showing an embossed Crowned eagle surrounded by a laurel wreath. In June 1920 the belt was replaced by one of woven silk in RAF blue, with gold embroidered edges. White gloves were also worn with this uniform.

Todays No1 uniform is used on ceremonial occasions in Home wear, with the addition of sword and aiguillettes for high ranking officers and their staff. Brown leather gloves are worn by Officers and Warrant Officers.

In warm weather areas No 6A Dress is the full ceremonial dress. A high necked single breasted white tunic, with six gilt (anodised) buttons and shoulder boards. Worn with white trousers, black shoes and standard blue/grey SD cap. Sword and aiguillettes are worn by staff officers.

MESS DRESS

As with the brother Services, many formal occasions in the Royal Air Force take place in the station messes. A more formal form of dress was required which was originally rigidly adhered too. During the 1939 - 1945 War, mess dress was discontinued by AMO A432 of October 1939, to return in 1947. On the return to use, a more flexible approach was allowed with the introduction of Interim Mess Dress, (AMO- A284), which is still used in today's Royal Air Force and known as No 4 Dress. All officers maintain a Mess Dress Uniform for formal occasions, Known as No 5 Dress and Senior NCO's retain the right to wear Mess Dress also.

Although announced by AMWO 291 of 25th March 1920, the details were not published until October of that year by AMWO 913, for the first RAF Mess Dress. The uniform was to consist of a high waisted blue grey, single breasted jacket, tapering to a shallow point at the front, below the waist. A wide stepped straight lapel, faced in B/G silk tapered to the waist, where the jacket was fastened by a metal loop attached to two gilt metal buttons. Three similar buttons, angled vertically, decorated the garment front on each side. Three smaller buttons extended from the cuff edge along the rear of the sleeve. The buttons were similar to Full Dress, having a separate gilt metal eagle and Crown attached to the domed,gilt metal button front. Rank was shown by half sized gold lace on shoulder boards. Above the lace a Crowned gold eagle was embroidered and above this a gilt metal button. Air officers had the button surrounded by gold embroidered laurel leafs.

Beneath the jacket a low pointed blue waistcoat was worn. Trousers were of the same material as the jacket and worn with black socks and black patent shoes. The white shirts worn at the time were starched fronted and had a detachable winged collar to hold a black bow tie.

This uniform was maintained for eight years, until the publication of AP (Air Publication) 1358, which illustrated orders of Royal Air Force dress and badges. The Mess Dress was altered by the introduction of overalls and Wellington boots to replace trousers and shoes as previously mentioned for Full Dress.

In December 1934 a complete redesign was published by AMOA311. This was to comprise of a new blue/grey jacket which had a shallower pointed front and lapels in ribbed silk. The front decorative buttons were more spaced out and angled from the lower front opening of the garment to almost the arm pit, three each side. The shoulder boards were abolished and rank was now shown by gold lace on the lower sleeve. A white, three buttoned waistcoat was also introduced at this time.

Overalls, as used in Full Dress uniform, were brought into use. These had the outer seams embellished with gold lace, with two bright blue silk stripes. Group Captains to Flight Lieutenants had 1 1/2 inch lace with 1/16 inch silk stripes and Air Commodores and above 1 1/2 inch lace with 1/8 inch stripes. Flying Officers and Pilot Officers retained the plain overalls.

Modern

In today's Royal Air Force, Mess Dress (No 5 Dress) follows closely the 1934 version. The cut of the jacket is similar. The lapels are now made in the same material as the rest of the jacket, with lower edges and one piece anodised aluminium buttons having replaced the three section ones. Trousers are plain and worn with black shoes and socks. White shirt with black bow tie are worn by all Ranks.

Officers

No 5 (A) Dress: As above with white waistcoat.

No 5 (B) Dress: As above with blue waistcoat.

No 8 (Tropical):   White jacket, cut as above. Shoulder

boards showing rank, blue/greytrousers.

Warrant Officers and Senior NCO's

No 5 Dress: Blue/grey jacket and trousers as above. No waistcoat worn, but blue cummerbund sometimes worn. Additional three buttons on lowersleeve.

No 4 Dress: Standard No 1 uniform worn with white shirt (Interim Dress)   and black bow tie.

WRAF

Mess Dress for lady officers had remained basically the same, with minor alterations since the 1950's. The 'tube' dress was RAF blue and covered the body from the loose turtle neck to the ankle length hem, with long sleeves. The only embellishment being an enamel brooch which indicated rank.

On the 12th June 1996, a new Mess Dress was promulgated for female officers. It consists of a jacket and long full skirt in B/G material. The jacket is almost a straight copy of the 1920-28 officers' Mess Dress, tailored to the female form, with the jacket front closed by a loop connecting two anodised buttons. Three buttons decorate each side of the long lapel on the tunic front. Officers rank is shown by cuff lace.

ROYAL AIR FORCE BANDS

Bands have been an important part of the fighting service since early times. A military parade or pageant is never complete without one. Musicians, like all trades, have to be taught and to this aim, the Royal Air Force followed the sister services by creating a school of music at

Hampstead in North London, on the 2nd July 1918. At first these were semi-official bands. The unit was then transferred to RAF Uxbridge, Middlesex to be reformed as the Royal Air Force Central Band on the 1st April 1920. RAF Uxbridge is still the home of music for the Royal Air Force and is now the location of Headquarters Music Services with the Central and the RAF Regiment bands.

In addition to the Central Band there exists three other established bands in the RAF. These being the Bands of the RAF College Cranwell, The RAF Regiment and the Western Band.

Due to the popularity of military music for parades etc, many Volunteer bands have formed at Royal Air Force stations, Auxiliary units and Cadet Forces.

All established bandsmen and bandswomen are trained medics and assist with casualties in the Aeromedical Evacuation role in time of war.

Originally bands of the RAF wore their standard Khaki Service Dress with the addition of white belts, slings and coloured dress cords. As the Royal Air Force uniform developed, the blue/grey No 1 uniform was used, with SD cap.

The established Bands have developed a distinctive B/G uniform of their own. Loosely based on the old Full Dress uniform, the tunic has a stand-up collar and is single breasted with seven anodised buttons. Decoration has varied over the years, but at the time of writing, a set of aiguillettes are attached to the epaulettes. The collar is edged with gilt cord for bandsmen and officers (Directors of Music) have gold oak leaves on the collar front. The trousers show a wide gold stripe along the outer seam, with a blue central stripe. Belts have been worn at various times throughout the bands' history and were blue grey cloth, edged with gold embroidered 1/2 inch tapes. An ornate belt buckle was worn, which is also reminiscent of the Full Dress uniform buckle. A two piece gilt buckle, with the left loop attached to a circular plate, around this plate has an engraved laurel wreath. In the centre,on a pitted background, a central crowned eagle and below the motto 'Per Ardua Ad Astra' are embossed. A flat hook, at the rear of the plate, attaches to the right hand belt loop. Directors of Music wear a gold sash belt with two parallel light blue lines or a sword, when on parade. Bandmasters (Warrant Officers), have a Light blue sash belt, or wear the officers sword.

No 9 Dress (Bandsmen/women).

All Bands wear a stiffened SD cap, bearing a gilt peak border for Bandsmen, with Directors of Music (Officers and Warrant Officers a Broad 3/4 inch (20mm) peak border. Directors of Music also have black piping around the crown of the hat. For Full Dress occasions,the established bands wear what must be the most impressive headgear in the RAF, a unique style busby.

Flat topped and oval in shape the busby is covered in black Canadian rabbit fur. The back sloping 3 inches below the front, to cover the neck. A gold plated link chin strap is attached inside to Porvar straps. On the front, an ornate RAF badge is worn (described in the badge section), which conceals a plume holder. Into this holder is placed a blue nylon, straight plume with gilt material base. Directors of Music (Bandmasters) wear a red plume.

Slung below the badge and attached to two RAF buttons at the sides, is a knotted golden dress cord, spotted with dark blue. Officers' and Warrant Officers' cords are plain gold.

From early 1996, a new uniform has been authorised for established bands when playing in concerts. Based on the current SNCO's Mess Dress the jacket is waisted and slopes to the front. Two anodised RAF buttons are attached at the waist by a metal loop, just below the long lapels. A loop on the shoulders allows the attachment of gold cord entwined knots. Trousers continue to have a broad, gilt stripe along the outer seam. The women's skirts are straight and ankle length. Badges are worn on both sleeves and the uniform is completed with black shoes.

Mess Stewards

Stewards are in attendance in the Officers' and Sergeants' Mess during the normal functioning of the Mess. They are especially visible at Dining-in nights, visits by Royalty and VIPs to Officers' Messes A

white single breasted jacket is worn, fastened by six anodised RAF buttons, and having a stand up collar. The shoulders are decorated with dark blue cord. Senior NCO's wear a dinner suit. On normal occasions rank is shown by the standard insignia or chevrons in chrome pinned to the sleeve. During Royal visits heavy gold shoulder cords are worn and WO badges and NCO chevrons are made of gold thread.

At the time of writing, catering at many Royal Air Force Stations has been taken over by civilian contractors, but it is hoped that the traditions and high standards will be retained.

CHAPTER V

BADGES 1920 – PRESENT

ROYAL AIR FORCE SERVICE DRESS

Badges of the Royal Air Force were developed originally either by copying or corrupting existing insignia of the sister services. The officer lace developed through the Royal Air Force of 1918 and has not changed since August 1919. However to keep our aircraft in the air requires many specialists, from air traffic control to messing. Air and ground crews needed to be trained to retain the safety and efficiency of the new aircraft types that developed rapidly as technology advanced. To this end trade badges were awarded to personnel to show achievement. These also added esprit de corps and aided recognition. The RAF has also taken on additional responsibilities over the years to fill developing needs. These have included rescue, airfield construction/defence as well as the comprehensive training.

The majority of badges in the Royal Air Force are of the cloth type, although some have developed from brass for economy reasons eg Physical Training Instructor. Cap and collar badges were made of gilded metal until 1953, when anodised aluminium (staybright) was introduced to replace them. These were intended to reduce the 'bull' of cleaning as they remained bright and any attempt to polish them resulted in the grey under surface showing. Although they undoubtedly reduced the cost to the MOD, they are brittle and are easily broken.

Cloth badges have also seen economies. During the Second World War many trade badges and shoulder eagles were printed on cotton cloth. Although functional, they were less attractive than embroidered badges and many clothing manufacturers supplied 'unofficial' embroidered copies, which were often bought privately to enhance the wearer's uniform. In most cases a 'blind eye' was shown by the wearer's unit and they were allowed to be worn.

Some badges were 'borrowed' from other nations, an example being the WAG half brevet, which derived from the Royal Canadian Air Force, but was not an official RAF badge. All Wireless Operators were trained Air Gunners and officially were to wear the Signallers' half wing.

CAP BADGES (OFFICERS)

Full Dress Helmet:
AMO 551 June 1921
A gilt circlet, originally a Garter, with above a King's Crown. Across the centre a siver eagle, wings extending the circle. Around the circlet *Per Ardua Ad Astra.*

Air Officers SD:
Cap Badge
AMWO 617 1918
Gold wire, embroidered laurel wreath bordering a black padded, shaped patch, broken at the top by a K/C or Q/C, with red cushion. Above the crown a crowned lion in gold wire. The centre has a gilt metal eagle with wings superimposed on the laurel wreath.

Air Officers F/S:
Cap Badge
AMO A 142 1940
As above 1/2 size. King or Queen's crown.

Service Dress cap badge:
AFM 2 1918
Centrally a gilt metal eagle
supported by four, gilt wire
laurel leafs. Above the eagle
a gilt embroidered crown
with a red cushion and white
base. All on a black padded,
shaped patch.

Beret badge:
1944
As above, 1/2 size.

Field Service:
cap badge
AMO A 93 1936
Separate gilt metal eagle and
crown. Spaced by backing
plate. 50mm across.

Cap Badges (Non Commissioned Officers & Other Ranks)

Service Cap badge:
AFM 2 1918
One piece, gilt metal or
Anodised. Flying eagle
supported by four laurel
sprigs and surmounted by
K/C or Q/C with red cloth
cushion (Some anodised
badges appear silver).

Warrant Officers:
beret badge
As above in gilt metal 1/2
size.

Warrant Officers:
Field Service cap badge
AMO A 93 1936
Crowned eagle, as worn by
officers

NCO's & Other Ranks:
Service Cap or Beret
(1919-1954) AMO 783/19
A laurel wreath, pierced at
the top by K/C, encircling
the RAF monogram in
gilding metal. Black plastic
badge used for beret during
later years of WWII.

NCO's & Other Ranks
(Nov 1954 -Present)
AMO 274/54
As above with Queen's
Crown. Gilding metal or
anodised aluminium.

March 1996
(SD Cap)
Cloth badge as above.
Monogram and Laurel
wreath embroidered in gilt
thread. Crown with red
cushion and coloured stones,
all on a black oval, padded
cloth patch.

Shoulder Badge & Titles

The shoulder eagle had continued on from the days of the early RAF and was only discontinued with the introduction of the woollen pullover uniform in 1973. As from February 1996 a new Eagle badge has returned to once more grace the No 1 Dress uniform. These badges are issued in pairs, the eagle 'flying' to the rear.

Home Service Dress
A flying eagle in light blue embroidered thread, on a B/G or black oblong patch. Issued printed, light grey on black during WWII.

Khaki Uniform
As above, eagle embroidered in red on Khaki Drill oblong linen patch.

Home Service
(1996)
A flying eagle in light blue and white embroidery with yellow beak. On B/G stiff oblong patch.

Shoulder Titles

During the 1939-45 War, the Royal Air Force received assistance from the Empire Nations, many of whose personnel served in the United Kingdom. Although the uniforms of Canada and New Zealand Air Forces were similar to the RAF, the Australian and South African uniforms were of a darker colour. Personnel also came from the Dominion Countries and were issued with Royal Air Force uniforms. To recognise the contribution given by these countries shoulder titles were worn to denote the wearer's country of origin.

These shoulder titles showed the name of the wearer's country embroidered in pale blue capitals on a black patch. The patch being curved for officers and rectangular for NCOs and Airmen. In tropical kit the titles appear as red on Khaki Drill. Forty-two were officially issued, although many unofficial ones are known to have been worn.

Below are listed the official Titles with their date of origin and the Air Ministry Order Number.

October 1940 AMO A760
Rhodesia

March 1941 AMO A219
Canada Australia New Zealand South Africa

August 1941 AMO A563
Newfoundland

December 1942 AMO A1303

Aden	Kenya
Bahamas	Leeward Islands
Barbados	Malaya
Basutoland	Malta
Bechuanaland	Mauritius
Bermuda	Nigeria
British Guiana	Nyasaland
British Honduras	Palestine
Ceylon	Seychelles
Cyprus	Sierra Leone
Falkland Islands	Somaliland Protectorate
Fiji	Swaziland
Gambia	Tanganyika
Gibraltar	Trinidad
Gold Coast	Uganda
Hong Kong	Windward Islands
Jamaica	Zanzibar

February 1945 AMO A163
India Burma

The wearing of above was abolished in April 1948 (AMO 368)

Many Servicemen had also escaped from their occupied countries and had been issued with RAF uniform. In some cases they had been allowed to customise their uniform with the addition of badges relating to their own air force. These badges do not fall into the scope of this volume, however shoulder titles with their country of origin were worn.

RANK BADGES

OFFICERS

As previously described officers rank is denoted by three sizes of braid. The braid is black with a central bright blue stripe and are 2 inches (50mm),1/2 inch (12mm) and 1/4 inch (6mm). This braid is cut to size from a roll for use with all uniform.

Raincoat
The 1972 pattern raincoat did not have epaulettes and rank was shown by means of a small oblong gilt metal brooch faced in D/blue enamel, which was worn on each lapel. These showed centrally a miniature example of the rank braid in the base gilt metal.

NON-COMMISSIONED OFFICERS

NCO's have remained the same as those promulgated for the blue grey uniform of 1918. Modifications of the Flight Sergeants' Crown and the inclusion of Leading Aircraftman/ woman (LAC) and Senior Aircraftman/woman (SAC) ranks being the main changes.

Warrant Officer:

Royal Coat of Arms in light blue on black or B/G shaped patch. Red crown cushion and animal mouths. King or Queen's Crown. Some patches cushioned.
Shoulder sides:
Dark blue 50mm wide material strip with Royal Coat of Arms in silver thread. Red crown cushion,animal mouths and part shield.
Wrist strap:
Brass or Anodised aluminium. K/Q crown Royal Coat of Arms. Worn on brown leather strap.

Flight Sergeant:
Three chevrons on a shaped patch with K/Q crown above. Two sizes known 6 inches(15cm),4 1/2 (11cm Crown originally large. Embroidered light blue on black patch reduced in size later to 35mm. Crowns used on No1 uniform in AA or gilded metal.

Shoulder slides:
D/blue, 50mm wide strip, chevrons and crown in silver thread.

Flight Sergeant:
Crown in light blue thread over three light blue chevrons.

Sergeant:
Three chevrons, as described above.

Corporal:
Two chevrons, as described above.

KHAKI DRILL UNIFORM

Flight Sergeant:
White square cloth patch used instead of crown over white tape chevrons.

Sergeant:
Three chevrons, as described above.

Corporal:
Two chevrons, as described above.

On the 1st of January 1951, the Senior Aircraftman/ woman title was introduced and together with Leading Aircraftman/woman became a ranking structure below NCO rank.

**Senior Aircraftman/
woman (SAC) 1950**
Three bladed propeller in light blue thread on a square D/blue or blue/grey patch. Worn with one blade pointed down

Shoulder slide:
Silver thread propeller on a 50mm wide D/blue slide.

Leading Aircraftman/ woman (LAC)
Twin bladed propeller in light blue, embroidered onto an oblong D/blue or blue/grey patch.

Shoulder slide:
Silver thread propeller on a 50mm wide D/blue slide.

Khaki Drill uniform:
Red propeller on a Khaki Drill oblong cotton patch.

The design of this badge has been used from the time of the Royal Flying Corps, when the appointment was called Air Mechanic 1st Class. Authorised in October 1916, it was re-introduced in 1938.

TECHNICIANS

In 1950 a reorganisation of the RAF's trade structure took place and two streams of promotion were created. To run parallel to the traditional Non Commissioned Ranks, personnel who's duties required a more specialised knowledge in communications and engines etc were termed Technicians and were distinguished by wearing their chevrons inverted. New titles were also used and were as follows.

Master Technician:
As RAF Warrant Officer

Chief Technician:
Three inverted chevrons with crown above.
Senior Technician:
Three inverted chevrons
Corporal Technician:
Two inverted chevrons
Junior Technician:
Single inverted chevron

The inverted chevrons continued for fourteen years until 1964, when the Technical ranks were abolished in favour of the traditional rank structure. Technicians regained their status by the introduction of a four bladed propeller badge,(AMO A63).

Chief Technician:
A small four bladed propeller in light blue thread,embroidered on a

round black patch. Worn above three chevrons.
Shoulder slide:
Silver thread propeller and crown on a 50mm wide D/blue slide.

Junior Technician:
55mm square black patch with a four bladed propeller in light blue thread.

Shoulder slide:
Silver thread four bladed propeller in light blue thread.

WOMEN'S ROYAL AIR FORCE & WOMEN'S AUXILIARY AIR FORCE

Members of the WAAF & WRAF wore/wear the same badges as their RAF colleagues. During the Second World War, female uniformed staff wore a letter 'A' to show their auxiliary status. In the case of Officers this was shown by a gilt metal 'A' on the lapel fronts. Other Ranks wore an embroidered 'A'in light blue thread on a square, B/G patch, which was worn beneath the shoulder eagle. In early photographs this eagle is shown on a patch which is trimmed to shape.

Since 1952, two sizes of Airwomen's cap badge have appeared in anodized aluminium, reflecting the change in head gear. Although the same design as the RAF's OR's badge, they are sewn onto the hat band and do not have fixing pins. One is of standard size, the other being of half size.

There was one different rank badge that was authorised by AMO

A212/40, that of Under Officer.

A large light blue embroidered crown, surrounded by an open laurel wreath, the crown having a red cushion. This rank being the equivalent of the RAF,s Warrant Officer insignia was worn on the lower tunic cuff. The badge was replaced in January of 1942 (AMO A104), by the WO's Royal coat of Arms. This order also authorised WAAF members to wear the RAF Flight Sergeant, Sergeant, Corporal and Leading Aircraft - women badges.

CHAPTER SIX

AIRCREW BADGES

The distinguishing 'Pilot Wings' have been worn, with slight modifications, since the creation of the Royal Flying Corps in 1912, and for a long time together with the Observers' half brevet were the only flying badges worn above the left tunic pocket. With the introduction of larger aircraft coming into service with the Royal Air Force, more crew members were required to operate them. The crew sizes reached their height during the Second World War, when large numbers of personnel were needed to crew bombers and Coastal Command aircraft.

The Half Brevet as we know it today originated from December 1939, when the Air Gunners brass winged bullet was changed to a cloth badge. Many of the classifications have gone out of use through time, as well as the title of the wearer, but at the time of writing, fifteen half brevets have been on the official list.

Described below, with the Pilots' Wings, are the officially issued half wings, with their dates and Air Ministry Order number. It should be noted that during WWII, Aircrew 'borrowed' half brevets of other countries were worn, the Royal Canadian Air Force Wireless Air Gunner (WAG) being a prime example.

PILOTS' WINGS

Pilots' Wings: *(1st pattern) Blue uniform AFM 2*
Outstretched wings 4 inches (100mm) from tip to tip. These support a laurel wreath encircling the RAF monogram with above a King's

crown with red cushion. Wings and wreath in gold silk with the monogram in silver thread. All on a black cloth shaped patch.

Pilots' Wings *Khaki uniform AFM 2*
As above, wings, crown and monogram in cream and wreath in bronze silk. The above Brevet has been used since 1920, with Crown variations, as the standard Pilots' Wings.

Pilots' Wings: *Blue uniform AMWO 1025 1918 to AMWO 153 1925*
As above, embroidered all in silver bullion.

Pilots' Wings: *Khaki uniform AMWO 783 July 1919*
As above, embroidered all in khaki silk, on a black shaped patch.

HALF BREVETS

The Observers' Half Brevet design had not changed since its introduction in 1918. During the twenties and early thirties the badge had fallen from use, but was reinstated by AMO A 347 of 1937, when the need to crew larger aircraft was recognised. Following the introduction of the Air Gunners' Half Brevet in 1939, all aircrew Half Wings followed the same design.

Observers' Half Brevet:
1918 (obsolete)
'O' in silver thread, with a gold wing extending to the right, as viewed, on a black patch.

Observers' Half Brevet:
AFM 2 1918 silk. and after 1920
(obsolete)
as above, embroidered in
cream

Air Gunners' Badge:
AMO 204 1923 (obsolete)
Brass winged vertical bullet.
Worn on upper right arm.

Air Gunners' Half Brevet:
AMO A 547 1939 (obsolete)
Two bronze embroidered
laurel branches, encircling
the capital letters AG,in
white,with a single wing
containing twelve feathers
extending to the right, as
viewed. On a black shaped
patch.

Radio Observer:
AMO A 402 1941
(obsolete)
As above, with the letters
'RO'.
The title changed latter to
'Observer Radio', Brevet
unchanged, and was used as
a security name for air radar
operators.

Air Bomber:
AMO A 1019 1942 (obsolete)
As above, with the letter
'B'.

Navigator :
AMO A 1019 1942 (obsolete)
As above, with the letter
'N'.

Flight Engineer:
AMO A 1019 1942 (obsolete)
As above, with the letter
'E'.

Wireless Operator (air):
*AMO A 1242 1943
(signaller) (obsolete)*
As above, with letter 'S'.

Meteorological Observer:
AMO A 409 1945
(obsolete)
as above, with the letter
'M'.

**Parachute Training
Instructor:**
AMO A 1079 1945
(honorary aircrew)
An open parachute in
white, replaces the letter.
Originally a circular B/G
patch showing a descend-
ing parachute within lau-
rel sprigs in white thread
was worn on an armband

Air Quartermaster:
DCI S 159 1950
(obsolete)
As Air Gunner,'AG'
replaced by letters 'QM'.

Redesignated to:

Load Master:
DC1 S 59 Sept 1970
As above, with letters
'LM'.

Air Electronics Operator or Officer:
AMO A 6 Jan 1963
As above, with 'AE'.

Fighter Controller:
1983
as above,with letters 'FC'.

A further Half Brevet briefly appeared during the 1960's, that of 'Air Technician', showing the initials 'AT'.

PATHFINDER BADGE

A distinctive badge worn by members of Bomber Command's Pathfinder squadrons, which came into force on the 15th August 1942. These highly skilled crews accurately located and illuminated targets ahead of the main bombing forces.

The badge was authorised in November 1942 (AMO A 1244) and showed a gilt metal RAF eagle, with a 2 inch wing span,which was worn on the left breast pocket flap. For security reason it was not allowed to be worn while on operational flying duties.

NCO AIRCREW RANK BADGES

Following the end of hostilities of the Second World War, a reorganisation of the Royal Air Force was started. Many aircrew were 'demobbed' to civilian life and a new training programme for professional aircrew was needed. From the 1st July 1946, ground trades and aircrew were divided into separate units, eventually even down to Mess level. Officer Aircrew retained their distinctive brevets, while Non- Commissioned Aircrew ranks were replaced by a new rank structure to equate to the Ground Trades.

The new badge for the NCO Aircrew was unofficially known as the 'Star & Garter' and was to be worn on both upper sleeves. It was a blue/grey cauldron shaped patch, bordered by laurel branches which were broken at the top by an eagle. In the centre a combination of stars denoted the rank. Master Aircrew differed by having the eagle centrally and a Royal coat of arms above. All detail was embroidered in light blue silk.

Aircrew I:
Three stars with crown above.

Aircrew II:
Three stars.

Aircrew III:
Two Stars.

Aircrew IV:
One star.

Aircrew Cadet:
No star. (Aircrew under training).

This move to segregate the different branches was not popular and was withdrawn in August 1950, with NCO Aircrew being reinstated with the normal chevrons.

Following the transition back to chevrons, Non-Commissioned Officers were to be distinguished by the addition of a gilt metal eagle worn above the stripes. One badge that has been retained, is Master Aircrew, which has been improved by the addition of a gilt metal eagle.

The above are worn on shoulder sides, with silver chevrons and gold eagle embroidered on D/blue.

A brass badge of the Master Aircrew was authorised to be worn on a leather wrist strap for tropical wear.

PRELIMINARY FLYING BADGES

Preliminary Flying Badges were a post war recruitment initiative designed to promote Aircrew training. Promulgated in AMO A 631 of September 1949, they were awarded to Aircrew trainees after the completion of basic training and worn on the left breast pocket, as viewed, until replaced by the appropriate Brevet or Half Brevet.

Pilot:
An RAF eagle, as the shoulder eagle, cut to shape, embroidered in L/blue thread on D/blue.

Aircrew:
A D/blue oval patch bordered by laurel branches with centrally the initial'N,S,E,AG',embroidered in L/blue thread. (Navigator, Signaller, Engineer, Air Gunner).

AIR STEWARD

Royal Air Force aircraft have been used to transport the Royal family to national venues for some years and with the increasing use of these aircraft for high ranking officers, VIP's and politicians, attention was required for a regular in flight service. For this purpose members of the catering branch were recruited to become Air Stewards, with a badge to be worn on the upper right sleeve, authorised on the 14th June 1967 (DCI S 119).

Air Steward:
The letters 'AS' contained in a laurel wreath broken at the top and supported by outstretched wings. All in light blue silk on a black shaped patch.

CHAPTER VII

DISTINCTION & TRADE BADGES

Chaplains

Chaplains in the Royal Air Force, in common with the other Services, all hold commissioned rank. Although addressed as 'Padre', they hold multi-denominational services, and wear SD uniform with a black cassock waistcoat and white clerical collar, for normal service dress.

Cap Badge:

A black cross pattée with centrally a gilt metal laurel wreath, containing the RAF monogram. The cross superimposed on gilt metal outstretched wings, with above, an embroidered gilt, K/Q crown with a red cushion. All on a black shaped padded patch.

Collar Badge:

As above without crown or patch, with a brooch pin.

Queens Honorary Chaplain:

(Clerical stole badge)
A gilt badge showing a Crowned Royal cypher without a laurel wreath.

CHAPLAINS ASSISTANT (WRAF)

Cap Badge:
Two rings, with an eagle superimposed across the centre, ensigned by a King's or Queen's crown. All in chrome.

Collar Badge:
A Latin cross in gilt metal.

Collar Badge : (RAF)
As Chaplains collar badge.

MEDICAL SERVICES

From the beginning of the Royal Air Force officers in the medical profession have been distinguished by the wearing of collar badges. In addition, a maroon cap band replaced the normal black one between July 1919 (AMWO 783) to 1920 (AMWO 571). Three medical and one Dental branch collar emblems have been worn by medical staff over the years.

Collar Badges

1st Pattern:
AFM 2 1918
A small gilt metal Caduceus of Mercury, (Two serpents entwined around a winged staff.)

2nd Pattern:
AMO 162 May 1918
A large (1 3/4 inch) gilt metal badge of a Crux Anasata, surrounded by a crowned laurel wreath, with below a scroll bearing 'NEC ASPERA TERRENT'. (Crux Anasata : A looped cross entwined by a serpent.)

3rd Pattern:
AMWO 571 1920 to present
A small crowned Caduceus of Mercury in gilded
metal or anodised aluminium.

FLYING DISTINCTION BADGES

Air Evacuation is now a common occurrence, to move casualties to better equipped hospitals behind the fighting areas. To assist in in-flight, long distance aid, personnel from the regular and Auxiliary RAF form an air wing and are distinguished by arm badges for doctors and orderlies. All ranks now wear the Caduceus of Mercury collar badge.

Flight Medical Officer:
1973
An embroidered
Caduceus of Mercury
with brown staff, gold
serpents and white
wings. All on a D/blue
fan shaped patch.

**Flight Nursing
Attendant:**

A red Geneva cross on a
L/blue field, supported
by white outstretched
wings. All embroidered
on a shaped D/blue
patch.

Dental Branch

Collar Badge:
AMO A 124 1942
A bronze or Staybright badge
showing a laurel wreath containing
the monogram 'DB', supported by
outstretched wings.

PRINCESS MARY'S ROYAL AIR FORCE NURSING SERVICE

Officers and Airmen/Women wear the same cap and rank badges as
other members of the RAF on the B/G No 1 uniform, the only additions
being the Caduceus of Mercury emblems
worn on the tunic lapels, these badges are
also shown on the lower opening corners of
the officers shoulder cape. A shoulder title
was worn by the OR's and showed the
letters 'PMRAFNS' in white on a black
curved patch.

Female members of the service, wearing
the white ward 2D Dress, show a badge on
the left shoulder front. Embroidered in blue
silk with white detail, it consists of an oval
ring on which is superimposed a crowned
Caduceus of Mercury, on a white oblong
patch.

A further badge was worn by trainee
nurses on their ward dress (No 2D), until
graduation to SEN standard. In December
1995, the badge ceased to be worn,
following the closure of the Royal Air Force
Nurse Training school. Nurse training now
takes place at either Portsmouth University
or the Royal Hospital Heslow, on a tri-
Service basis.

Student Nurse Badge:

A 1 1/4 inch L/blue enamel disc with a gilt metal border, and the wording 'NURSE TRAINING SCHOOL * ROYAL AIR FORCE' embossed in L/blue on it's circumference. In the centre a raised gilt metal laurel wreath and Caduceus of Mercury. The badge has a brooch pin fixing with safety catch.

RESCUE SERVICES
Air Sea Rescue

With the increasing combat activity over the English channel during and after the Battle of Britain and the larger bomber formations now being used over Germany, many aircrew were forced to bale out or ditch in the sea. Patrols of Coastal Command Walrus aircraft had been used to rescue these valuable aircrews before death by exposure or capture by enemy vessels. As these incidents increased a more practical solution was sought,the result was the high speed wooden launches, powered by Thornycroft engines. Originally these vessels where unarmed, but following attacks by enemy E-boats and aircraft, padded applique armour and turret machine guns were installed.

These vessels were manned by members of the RAF Air Sea Rescue service, which was formed in February 1942. A distinguishing badge was eventually promulgated by AMO A 17 of 1943, to be worn on the upper right sleeve.

The badge was printed on a black square patch and showed a powered launch. Within the slung aerial wire are the letters 'ASR', all detailed in off white.

The use of this badge was abolished on the 1st June 1948 by AMO 368. The nautical branch of the Royal Air Force was finally disbanded on the 1st April 1986.

Mountain Rescue

Aircrew also have to be rescued from inaccessible areas on land and in the UK, these are usually in the mountainous regions of Scotland

and Wales. Civilian mountain rescue team volunteers have been available to assist the civil authorities in these locations, for the recovery of aircrew as well as climbers and walkers. Teams of airman

volunteers from local airfields often assisted in these rescues and on the 23rd February 1959 AMO 38 acknowledged the work of these RAF volunteers by issuing a suitable badge.

A circular dark blue patch with centrally a coil of rope entwining two crossed ice axes, in L/blue silk. The word 'Mountain' appears at the top and beneath the rope the word 'Rescue', all writing in white silk.

A Rescue Volunteer badge has recently been authorised by Dress Regulations, to be worn by RAF Service personnel who are not part of the Mountain Rescue team but will assist when required. The badge shows the word 'RESCUE' above the curved word 'VOLUNTEER', with beneath two hands grasping each wrist. Detail in L/blue cotton on an oblong D/blue patch.

Telecommunications

The first 'trade' badge to be authorised in the Royal Air Force was on the 19th September 1918 (AMO 1066), that of Wireless Operator. This order states that the badge will be worn below the bird badge,(eagle) on both arms by qualified wireless operators and would be supplied in red worsted embroidery.The design continues from this time, where today it is worn on the No1 uniform by all qualified ground crew below the rank of Warrant Officer, who are employed in the radio engineering ground trades.

Centrally, a right fist clutching six lightning flashes in light blue silk on a black or D/blue oblong patch. Printed version, off white on black patch.

Shoulder slide:
D/blue embroidered detail on a stone (Desert Storm) background.

Physical Training Instructor

Fitness has always been of prime importance in the Armed Services and following reorganisation after 1918, a brass badge was promulgated for the use of fitness instructors in 1923. This badge continued in use until October 1949, although economies dictated that a cloth version was introduced during WWII. From this date a redesigned cloth badge was to be used.

Early badge:
AMWO 2O4 1923-1949
Three bent arms holding Indian clubs, forming a circular shape. Within a central circular plaque, raised capital letters'PTI' In brass stamping or L/blue silk on a D/blue, round cloth patch.

1949 - present:
Crossed swords superimposed by a RAF eagle, surmounted by a crown, with a red cushion. The K/C version appeared on a black or B/G square patch, with the Q/C version embroidered on a D/blue square background.

Singlet Badge:
(present)
The crossed sword badge in light blue silk with red crown cushion, on a white 95mm cloth square.

Parachute Jumping Instructor

When parachuting was in its infancy, it fell to members of the Physical Training branch to train infantry volunteers to jump safely. Following their success in developing jumping techniques and air dispatch duties, a badge was authorised to be worn on the upper arm.

The badge, on a black circular patch, depicts an open parachute in off white silk, with a laurel branch on each side in L/blue or off white thread. Nicknamed the 'light bulb' it was in use between 1943-45 when it was replaced by the half brevet. (AMO A 467 May 1943).

Bomb Disposal

In January 1941, a badge was authorised by AMO A 69, to be awarded to armoury technicians who were employed in making safe and the removal of unexploded bombs dropped on RAF airfields during Luftwaffe raids.

Worn on the upper right sleeve, it showed a descending bomb flanked by the Capital letters 'BD', surrounded by a laurel wreath, all embroidered in L/blue on a circular black patch.

Following hostilities, the badge was abolished on the 1st June 1948 (AMO 368), although it can still be seen on armlets of personnel employed in the disposal of ordinance, being re-authorised by (AMO A 92) of the 27th January 1949.

The author has also seen this badge sewn on the working blue pullover fore arm.

RAF Marksman

A high standard of marksmanship has always been the 'hallmark' of the British Armed Services, with competition between the services

keenly fought. To "encourage rifle shooting and foster a higher standard of marksmanship in the RAF" AMO 570 of 11th August 1949, authorised a badge to be worn on the lower left sleeve, which was to be re-qualified for annually. Still worn today, the badge shows crossed (Lee Enfield .303) rifles, embroidered in L/blue cotton on either a B/grey or black shaped patch.

ROYAL AIR FORCE BANDS

All musicians of established bands of the RAF are distinguished by the wearing of a metal lyre badge with crown on both upper sleeves of their No1 dress uniform.

The badge shows a five stringed, crowned lyre on a stand, with each lower side being cupped by oak sprigs. The badge is attached by two eyes with split pin and backing plate. Originally made of brass, in the K/C version, the latter Q/C badge is of anodised aluminium.

As previously described, established bands wear a distinctive uniform when playing on formal occasions. The badges worn on the tunic are also unique, as they are made of gold tape and embroidered in gold thread or metallised polyester.

Busby badge:
The RAF monogram, surrounded by a laurel wreath, broken at the top by a Q/C. The crown showing a red cushion, coloured 'gem stones', and a white base. Silver 'dots' cover the crown top with the remaining detail in gold thread. This badge is embroidered on a 3 1/8 inch (80mm) black padded, shaped patch. The Bandmasters' badge is in gold wire.

The above badge has been adopted in a smaller size for standard RAF use on the SD cap.(1996),but made of metallised polyester. The following badge descriptions refer to No 9 dress band Uniform.

Collar badge:
An RAF eagle, supplied in pairs, to allow each to 'fly' rearwards, with a Q/C with red cushion, silver top and coloured 'gem' stones, between the wings. The eagle in gilt thread, embroidered on a D/blue shaped patch.

Shoulder title:
The Central Band and the RAF Regiment Band wear shoulder titles
bearing their name in gilt wire capital letters on a curved D/blue
background. The Central Band is without a border and the Regiment
a copy of their standard shoulder title.

Rank badges:
Technician, Jnr Tech and SAC are in gold
wire or thread on a D/blue patch.

Musicians badge:
A gold thread lyre and base with five silver strings, acorns and oak leafs extend from the lyre base. Above a padded Q/C, embroidered with a red crown, coloured 'gem' stones and silver top. All on a D/blue shaped patch.

Band Master:
A larger version of above.

Warrant Officer:
Royal coat of arms in gilt thread, crown as above, worn above the Bandmasters' badge.

Chevrons:
Chevrons are made of gold metallised polyester and are 3.5 inches (8.8cm) wide.

Drum Major:
Four inverted gold chevrons with drum badge above. The side drum is in flat gold wire with gilt cord 'ropes' and rests on a base in gold thread. Above the drum a 'balloon' shaped pad, also in gold thread. The drum is on a D/blue,'Tomb stone' shaped patch.

Drum and chevrons are worn on both lower sleeves. Chief Technicians wear a small four bladed propeller above the

drum and flight sergeants a crown above the drum.

VOLUNTEER BANDS

Volunteer bands wore the lyre badge as well, but without the crown. (AMWO 848 20th November 1924).

Bugle badge:
The crossed bugle badge is found in brass and and was worn in the early days of the Royal Air Force, when orders were relayed around a camp by bugle calls.

Today RAF volunteer bands have three badges available for wear, these are in chrome metal and show a side drum, crossed bugles or a Crownless lyre.

ROYAL AIR FORCE REGIMENT

Defence of air force airfields and installations was originally the responsibility of local Army units, but with the additional threat of air attacks and the Army's increasing commitments, ground crews began to defend their bases with machine guns that had become available.

Commanded by attached Army officers, RAF personnel became dedicated to the ground defence role and by 1940 a badge was authorised to denote the 'Ground Gunner'. These gunners were still only armed with machine guns, with the Army controlling the anti-aircraft guns.

With the many airfields in operation during World War II, at home and abroad, and the effective way that Germany was using it's airborne forces at the time, a parliamentary committee recommended that the Royal Air Force should have a ground defence force under the control of the Air Ministry. To this end, the RAF Regiment was formed on the 1st February 1942, which absorbed all existing Ground Defence Flights then in existence and took responsibility for all forms of airfield defence.

Although maintaining their service dress 'blues', Regiment personal have worn the army service dress of the time, Khaki Heavy Duty dress,

denims and tropical kit to todays Nato combat dress. Badges worn on Heavy Duty dress and denims during World War II were all printed. Some badges are unique to the Regiment and are as follows.

Service Dress

Shoulder Title:
Home service
(AMO A 466)
1942
A curved D/blue / B/G patch, to follow the shoulder seam, with 'RAF REGIMENT' in capital letters and border in L/blue. Embroidered or printed.

Tropical Dress:
(AMO A 963) 1942
As above, red detail on a Khaki drill patch.

Desert Storm:
Dark blue embroidered border and letters on sand background.

Shoulder Slides:
(1990)
These cover the epaulette and show at the bottom end the letters 'RAF'centrally above the word 'REGIMENT' in black thread. D/blue L/blue and stone slides are available.

Parachute Wings

The Royal Air Force retains a parachute trained force in No 2 Squadron RAF Regiment. Formed in July 1962, the unit has been deployed in Cyprus, Aden as well as other international trouble spots. The qualification badge is worn on the upper right sleeve and consists of a white descending parachute supported by L/blue outstretched wings, embroidered on a D/blue or black shaped patch. Personnel attending and passing the parachute training course, although not part of an airborne unit, wore a circular D/blue or black patch showing a descending parachute in off white thread.

Signaller:

(AMO A 207) 1944
(obsolete)
A circular D/blue patch showing crossed flags with four lightning bolts emanating from the flag staff junction. All detail in L/blue.

Ground Gunner

Although not strictly a RAF Regiment badge I will include this badge in this section.

Ground Gunner:

1940 (AMO A 761) to 1942 (AMO A 466)
A D/blue oval patch with two large 'G's, with beneath two laurel sprigs.

CHAPTER VIII

MISCELLANEOUS BADGES

Works & Buildings

With the expansion of the RAF, following the Great War, Airmen were being employed in the construction and deployment of many new airfields and installations. To recognise the airmen involved in this work a badge was authorised by AMO 825 of the 20th November 1921, which was worn until it's withdrawal in 1929.

Cap Badge:

A brass badge showing a Masons square forming a 'V' shape, with the letters 'W&B' wedged within the angle, and above a King's Crowned eagle with outstretched wings touching the square's tips.

Collar Badge:

As above, 1/4 size.

Apprentices and Boy Entrants

Following the 1914-18 conflict, the need to recruit permanent skilled Air Force personnel was realised by Lord Trenchard and to assist in training the apprentice system was established on the 17th April 1919.

RAF Halton was one of the first schools for technical training, together with RAF Cranwell. Later schools were established at St Athan, Compton Basset and Yatley Boys, as well as a radio school at RAF Locking.

The Apprentices and Boy Entrants wore airmen's uniform with the SD cap, the black mohair band being replaced by a coloured one. Seven coloured bands were used to denote apprentices at Halton and Cranwell with Boy Entrants wearing brown, (AMO A 770 September

1947). In July 1950 this system was modified to diced coloured bands, each course and school maintaining it's own colour scheme.

Below are the cap band colours promulgated in AMO N 708 of July 1950.

No 1 Radio School

RAF Locking	No 1 Wing	Royal blue & Green
	No 2 Wing	Orange & Red

No 2 Radio School

Yatesbury School   Black & Scarlet

No 3 Radio School

Compton Bassett   Brown & White

No 4 S of TT

RAF St Athan   Red & White

In addition to the cap band, apprentices showed various coloured discs behind their RAF cap badges to denote their course.

Other distinctions of the uniform were the apprentice badge above half sized chevrons, when worn. (AMWO 204 12th April 1923).

Apprentice Badge:

A four bladed propeller enclosed within a ring, all in brass, or more rarely anodized aluminum No fixing pins or tang. Sewn directly onto the uniform sleeve.

Officer Cadets
RAF Cranwell

Royal Air Force Cranwell was one of the forerunners of aviation history being opened on the 1st April 1916 as a Royal Naval Air Service station for training personnel in the art of power flying and ballooning. On the 1st April 1918 the station became part of the newly formed

Royal Air Force. Flying training continued, as well as courses for Boys and radio operators.

A site was chosen close to the balloon section, to erect a purpose built Training College, in grand style, which was officially opened by the Prince of Wales on the 11th October 1934 to house 150 cadets. Today RAF Cranwell is a single gate entry for potential Officers of the Royal Air Force and is increasingly absorbing flying squadrons and administration duties.

Following initial officer training of six weeks, the Officer Cadets are streamed into courses for their intended future roles, pilot, navigators, engineers etc. Standard RAF uniform is worn, with a white cap band replacing the black mohair one. A white disc is also worn behind the Cadets' beret badge. These alterations date back to AMO A 375 of the 5th May 1947. Progression through training is shown by the wearing of shoulder slides on No 2 uniform and pointed tabs on the upper step of the No 1 tunic.

Initial six weeks' basic course: White slides with a central grey
. ribbon.

No 1 uniform on streaming to vocation course:
65mm x 27mm white tab with one pointed end and a central, longitude 3mm cord in the course colour.

No 2 uniform: White slides with central 13mm ribbon in course colour.

No 1 uniform following initial exam passes:
65mm x 27mm white tab with one pointed end, with the central course colour cord. Tab edged by 3mm yellow tape with a RAF button sewn to the pointed end.

These are tacked to the uniform lapels and ripped off unceremoniously following graduation.

No 2 uniform: White slides with a central 13mm course colour ribbon edged by yellow tape with centrally, a RAF button.

RAF POLICE

The Royal Air Force Police have worn the RAF uniform of the day, with the addition of white webbing. In 1953, AMO A 315 authorised the wearing of a white belt, cross-piece, pistol holster and gaiters, to be worn with the War Service Dress. These instructions also re-introduced the white topped SD cap. An arm band was worn on the right sleeve and was black with a wide red central stripe on which the word 'POLICE' appeared.

A similar arm band is currently worn on the No1 uniform, which is made of a synthetic felt and has a four double press-stud fastening, attached centrally above the letters is a crowned gilt metal eagle. A further arm band, which is divided equally with black border and central red stripe, and made of cotton, has the words 'RAF' on the upper border and 'POLICE'across the lower border. Centrally between these appear a corporals stripes. All detail in white.

With the No 2 uniform jersey and shirt now being worn, shoulder slides are used more frequently. These follow the theme of the arm

bands, being black with a central red stripe. The more common has the initials 'RAFP' in black along the red centre, although a similar slide is available for Assistant Provost Marshals with the letters 'APM'.

COMBINED OPERATIONS

Worn by members of the 3207 RAF Service Command and Communication Sections. The Service Command moved forward with advance units of the Army and prepared temporary airstrips and refuelling areas for operational use, before being relieved by better equipped construction crews. The job of the Communication Sections was to co-ordinate wireless traffic between the Services and therefore provide local air defence and attack aircraft when needed by the Army or Navy.

Combined Operations badge:
AMO A 1186 1942
AMO A 536 1946

The badge was supplied in mirrored pairs and consisted of an upright anchor to superimposed by a Thompson sub-machine gun with above, the RAF eagle. Printed in red on a black tomb stone or round shaped patch.

EAGLE SQUADRON

71 Squadron was the first RAF squadron during the Second World War, to be manned by American volunteer pilots. To mark their contribution, a badge was authorised to be worn by officers on their upper tunic arms by AMO A 818 of October 1940.

The badge showed an American eagle with above the letters ES and was embroidered in L/grey silk on a dark grey patch.

Judge Advocate

Worn by Judge Advocates under service conditions as a shoulder slide, the badge, in blue and gold, was promulgated in December of 1950 (AMO A 804) and shows a heraldic crowned shield, containing centrally the letters 'JAG', with above crossed swords. Below the letters the RAF eagle. The shield is supported by crossed Lord Chancellor's maces and bordered by laurel branches.

Mess Dress

Officers continue to show their rank by cuff rings in gold lace, in the style of the 1934 Gala Mess dress. Senior Non-Commissioned Officers' rank badges are worn on the right sleeve only in the same position as the No1 uniform. Warrant Officers and the Flight Sergeants' badges are embroidered in gold and silver wire, with red detail,with the Chief Technician's propeller in gold wire. Chevrons are in gold lace.

WRAF Officers, NCO's & PMRAFNS

Mess Dress Rank Brooches

As already mentioned in the uniform section, prior to 1996 female officers' rank was shown on their Mess Dress by means of an enamalled brooch. This was oval in shape and made of gilt metal. A laurel wreath, open at the top supported crowned wings, with the centre of the wreath in dark blue enamel. Across the centre, a gilt metal, horizontal bar or bars of different thickness would denote the officer's rank.

Senior NCO rank badges are made in chromed metal and attached by twin screwed posts, onto a backing plate. The following have been authorised in Dress regulations.

Warrant officer   Master Aircrew
Flight Sergeant   Chief Technician
Sergeant

The Chief Technicians badge was a copy of the original apprentice badge. Also available in a gilt metal brooch were the half brevets for Air Loadmasters and Air Quatermasters, as well as Dental and medical badges.

On the 1996 Mess Dress, the NCO's rank is shown on the right sleeve of the jacket in the same manner and style as their male colleagues.

Half size pilots' brevets and half brevets are available for all authorised aircrew categories. They are worn above the left breast pocket and are embroidered in gold on a B/G shaped cloth patch.

Good Conduct Stripes

Introduced before WWII, they denoted length of service and were either a single inverted stripe to show three years service or two stripes to indicate eight years. Worn on the left lower sleeve, they were abolished by AMO A 594 in 1950.

JOINT ARMS CONTROL IMPLEMENTATION GROUP

The Joint Arms Control Implementation Group (JACIG) was formed on the 1st August 1990. It is a tri-Service organisation tasked with the implementation of a number of international arms control treaties and carries out inspections of foreign military equipment abroad to confirm that arms reductions, agreed by the treaties, are being carried out. The group is also responsible for escorting foreign inspection teams arriving in the UK.

Although the group is mainly staffed by Army personnel, with members of the Royal Navy, Royal Marines and civilians, 36% are from the Royal Air Force and therefore the badge is justifiably included in this text.

The cloth badge shows a simplified Union Jack flag in full colour, with beneath the initials 'JACIG' embroidered in white silk. These appear on a violet, felt, square patch.

CHAPTER IX

AUXILIARY & RESERVE FORCES

AUXILIARY AIR FORCE

Following the armistice in November 1918 and the inevitable reduction (295,000 to 29,000 personnel) in the Royal Air Force, Sir Hugh Trenchard had no wish to see the decline of the newly formed Air Force. Regular personnel numbers would be reduced with the formation of a volunteer force based on territorial principles and would be an advantage in times of crisis.

In October of 1924, final approval was granted and the Auxiliary Air Force (AAF) was formed. This force was to relieve regular personnel for service abroad in times of conflict and was a 'Home Defence' force with commitment to UK service only. Squadrons were formed from the local populace and centred on a County basis, with meetings and training carried out at local airfields and town centres. AAF members were required to attend weekend and evening parades for training and hands on duties to ensure the smooth running of their squadrons.

With the deterioration of the international situation and the rearmament of Germany, the threat of a European war increased and the AAF went though a number of progressive changes to enlarge. By the 3rd September 1939 the Auxiliary Air Force manned twenty squadrons with a mix of fighters, bombers and Army Co-Operation aircraft, but recruitment had not kept up with this expansion and it was found that one third of the squadrons strength was made up of regular Airmen, whose presence was still required to maintain efficiency.

With defence of the British Isles still the primary duty of the AAF, the operation of balloon barrages was entrusted to them in may 1938. Officers were appointed for a four year period with Airmen required for a five year enlistment. As in the rest of the AAF these personal had only 'Home Duties' requirement and initially age limits for recruitment were 32-50 years for officers and 38-50 for Airmen. As the War progressed women, who had enlisted into many sections in the RAF, were used in increasing numbers. When War broke out, the balloon section of the AAF had a complement of approximately 16,400 Officers and Men/Women.

As already mentioned, AAF personnel had commitment only to their unit and their enlistment agreement ensured that they remained in the United Kingdom. With the pressures looming, the Royal Air Force mobilised on the 24th August 1939 and following the Declaration of War on the 3rd September 1939 an Emergency Act was passed which changed AAF personnel status by stating that Officers or Airmen joining after the 26th April 1939, were liable for overseas service unless they were in a Reserved Occupation.

The Auxiliary Air Force served throughout the War years with distinction, along side the regular Air Force, with their squadrons scoring notable success against enemy aircraft as well as downing a number of V1 rockets.

With the cessation of hostilities in Europe, the AAF was disbanded in 1945, only to have the flying squadrons reinstated in June 1946, with a Royal added to their title on the 16th December 1947 the AAF became the RAuxAF

During the post war years, flying squadrons continued their Home Defence role with the later marks of Spitfire and Mosquitos, as well as progressing through Meteors and Vampires. Other areas of defence were also manned by the RAuxAF to include Radar Reporting, Maintenance Support, Intelligence as well as additions to the Regiment strength in the form of Light Anti-aircraft Batteries. These units were to continue until 1957, when the flying squadrons were disbanded. A combination of advancing technology and defence spending cutbacks caused their demise. By 1961 the only RAuxAF units left in service were in Maritime H/Q posts.

Following the Russian involvement in Afghanistan in 1979, a decision was made to resurrect the RAuxAF Regiment for local airfield defence. Initially four regiments were experimentally formed and following their success two more followed in 1982. Additional manpower was also found to be needed in Aeromedical Evacuation and Logistic fields and these were formed in the same year.

ROYAL AIR FORCE VOLUNTEER RESERVE

In 1936, an additional reserve force was founded to run parallel with the Auxiliary Air Force, although to be more integrated with the RAF. This force, the Royal Air Force Volunteer Reserve, was an officer only reserve, designed originally to train pilots. The personnel were 'individual' reservists as opposed to 'unit' reservists, as in the AAF, allowing a more flexible approach to posting in time of conflict.

The RAFVR was open to all classes and commissioned promotion was awarded on completion of training, following an original intake as Airmen Pilots,to those of twenty one years or above. Training was based at civilian aerodromes and town Headquarters were to provide additional ground instruction. These Headquarters were also to be used to form a Ground Section of the RAFVR, with the training of Officers for Administration, Signals, Engineering, Intelligence etc., and in October 1937 the first of these was formed, a Medical Branch. An Equipment Branch was formed in January 1938, to be followed latter by Administration and Special Duties in that year. In 1939 Meteorology and Dental Sections were added to the duties in which the RAFVR were to assist the RAF in the coming conflict.

1938 also saw an expansion of the RAFVR Flying Branch, as the ever pressing needs of the flying services became apparent and the original forecast of 2,400 pilots was increased to 7,000. In addition Observers, Wireless Operators and a thousand Air-Gunners were required.

By September of 1939, the RAFVR had been unable to reach full efficiency due to financial restraints and diversification of adequate numbers of man power, however when War was declared of the 12,600 Reserve Officers available, 3,000 were from the RAFVR and AAF. During hostilities, conscripted Officers entering the RAF, held a commission in the RAFVR for the 'Duration of Emergency', so as not to interrupt the promotion prospects of career Royal Air Force Officers.

World War II ended with the RAFVR aircrews distinguishing themselves alongside their RAF colleagues and gaining many decorations including Victoria Crosses.

As with the AAF, the RAFVR was disbanded in 1945, only to be reconstituted in 1946, when duties were to assist in administration and training National Servicemen during their compulsory reserve service. Some new flights were created during the Korean Conflict to assist in fighter control and intelligence, the latter was the main survivor of successive defence cuts.

On the 1st April 1997, the Royal Air Force Volunteer Reserve was amalgamated with the Royal Auxiliary Air Force, following 60 years service,and the remaining RAFVR Officers, who are employed in Public Relations, Translation and Photo Intelligence, will now be on the strength of the RAuxAF.

ROYAL AIR FORCE VOLUNTEER RESERVE (TRAINING)

The RAFVR and AAF have within their remit a 'War-Appointable' role in times of conflict; the third element of the reserve forces is the Training Branch.

The RAFVRT is an Officer only reserve and is not war appointable. It is responsible for the running and instruction of the Air Training Corps and the Air Force element of the Combined Cadet Force.

More information in the Cadet Force chapter.

Badges

Members of the Reserve and Auxiliary Services wear/wore the uniform of the Royal Air Force of the time,rank and trade badges being the same. To distinguish the Reserve from regular Air Force personnel lapel badges were worn by Officers on the upper step of the No1 uniform or later on the shoulder straps. The lapel badge was in two sizes, 3/8 inch for the No1 uniform and 1/2 inch for the greatcoat shoulder straps. Authorised by AMO 378 of the 14th September 1939, they consisted of the capital letters'VR' for the Reserve and 'A' for Auxiliary. They are made of gilt metal or brass until the introduction of anodised aluminium. The same Order instructed Other Ranks to display a black square 1 inch cloth patch with the relevant initials in L/blue, to be worn on both upper sleeves, below the eagle.

Following the amalgamation of the RAuxAF and the VR, all officers will wear the anodised letter 'A'.

Members of the Volunteer Reserve and Auxiliary Air Force originally were entitled to wear the silver initials on civilian clothes by AMO 265 of July 1939. A chrome RAuxAF lapel badge is still available for civilian dress. This badge consists of a circle with raised edges. The top surmounted by a Q/crown and the bottom of the circle superimposed by a RAF eagle. Within the open circle a bar carries the letters RAuxAF.

An additional lapel badge was worn, above the VR, by the officers of the Education Branch and was promulgated in 1940 by AMWO A 116. It showed two crossed flaming torches, with a RAF eagle above and was made of gilded metal. This badge was worn throughout the War years and was finally abolished on the 29th April 1948 (AMO 368).

Chapter X

CIVIL AIR GUARD, UNIVERSITY AIR SQUADRONS, AIR TRANSPORT AUXILIARY, FERRY COMMAND

CIVIL AIR GUARD

With European tension mounting in the late 30's and the 'Reserve' build up not reaching the projected manpower levels, ways were sought to enlist civilian flying organisations to assist in training and to utilise qualified pilots who were not within the enlistment criteria.

Interest in aviation during the 1930's had become very popular, but only the well heeled could afford flying lessons, so it was proposed in the early part of 1938 to form an organisation based around civilian flying clubs. Here men and women between the ages of 18 and 50 could train to become pilots at a reduced cost. This would then be an additional source of partly trained pilots and also emergency facilities, should they be required.

The organisation was to be known as the Civil Air Guard and following the opening in July 1938, the enthusiastic response threatened to swamp the training facilities and instructors available, with 36,000 applicants applying by the end of 1938. As already mentioned, CAG members were to receive reduced tuition fees, £2 subsidy per head, (half price), to bring them up to the Air Ministry Amateur Pilots licence 'A' standard. In return they made the 'honourable undertaking' to serve in the Royal Air Force Reserve in times of 'Emergency' and being entirely voluntary, no bounty was paid.

The Civil Air Guard was essentially run at club level. The Air Ministry paid grants to assist in financing the venture, although co-ordination was to be undertaken by a Commission Chairman in the person of Lord Londonderry, a retired Secretary of State for Air.

Being a civilian flying organisation, uniform, as such, consisted of dark blue overalls and a field service cap of the same colour, usually privately purchased, and no additional uniform was worn. Badges were supplied to enhance this basic uniform and were as follows.

BADGES

Cap Badge:
A chrome laurel wreath pierced at the top by a K/C, with across the central void the angular capital letters 'CAG'in blue enamel. The 'A' being superimposed on the other letters.

Arm Badge:
A miniature embroidered ensign, showing a D/blue St George's cross with a white border, on a L/blue ground. In the first quarter a Union flag in full colour.

Pilots Wings:
Centrally the letters 'CAG' embroidered in D/blue with a white border, supported by angular white wings with four feathers. All on a shaped 120mm, L/blue Calico patch.

Buttons:
These were domed with the letters 'CAG'raised from the surface and in a pitted finnish. The button was made of brass with a chrome front.

Civilian lapel badge:
Two types appear to have been worn. A small copy of the pilots' wings in chrome with D/blue enamel letters and a 'button' type, with the letters 'CAG' in D/blue enamel on a pitted background.

UNIVERSITY AIR SQUADRONS

A further source of flying training was in the form of the famous Universities of Britain, who in the spirit of 'airmindedness' of the decade, had created their own flying 'Squadrons' as a week end, evening activity for their students. The first squadrons to be formed were at Oxford and Cambridge as early as October 1925, to be followed in latter years by other Universities. Apart from the flying training, two other flights were formed to work on engineering development and technical subjects. This formula follows on today, although training now includes electronics and intelligence and is more akin to officer training.

University Squadrons were closed down at the beginning of hostilities in 1939, but were re-opened in October 1940 as pre-entry non flying units for Officer Training.

Following World War II, the UAS suffered the general cutbacks that followed the cessation of hostilities. Sixteen squadrons survived from the 1940 total of twenty three. These squadrons struggled on, encouraging students towards the Royal Air Force, until the introduction of the University Cadetship scheme in 1968. Renewed recruitment interest and sponsorship by the RAF has allowed these 16 squadrons to enlarge to a student complement of 725 instructed by 76 qualified staff.

There are three forms of entry into the University Air Squadrons today. Predominantly they are RAFVR (now Auxiliary) Student Pilots with Officer Cadet rank. Other UAS students have already undergone preliminary officer training at RAF Cranwell and having passed Aircrew Selection, continue with their university studies as acting Pilot Officers. Thirdly a bursary can be applied for, whereby a grant is awarded to the student before university entry. These bursaries are awarded to Ground Trades as well as Aircrew and on completion of the Initial Officer Course, they are enlisted as Officer Cadets in the RAFVR (now Auxiliary).

Uniforms for students have improved since the 1930's, when it reflected the civilian character of the UAS, as it consisted of a dark blue blazer, showing a RAF crest badge on the breast pocket and was complimented by a squadron tie. This state of dress continued up to the temporary disbandment in 1939. During the war years as a unit for pre-entry training, RAF uniform was worn and this has continued up to today.

BADGES

Officers are commissioned in the RAF VRT and apart from the VRT collar badges wear standard Royal Air Force uniform and rank lace.

STUDENTS
Cap Badge:
Other Ranks anodised RAF monogram badge worn on a white circular background.

No 1 uniform:
White square tab with a D/blue stripe. (worn on the upper lapel step)

Arm Badge:
An oblong D/blue patch,with a curved top. An inner border containing at the top the University name, with below 'UNIVERSITIES AIR SQUADRON'. All embroidered in L/blue silk.

No 2 Uniform:
(A Students)
White square cloth patch worn on shoulder epaulettes.

No 2 Uniform:
(Bursary Students D/blue stripe. Worn on shoulder)
White square cloth patch with dark blue stripe. Worn on shoulder epaulettes.

No 2 Uniform:
RAF Pilot Officers' rank braid.

AIR TRANSPORT AUXILIARY

With the impending European war looming, scheduled civilian fights by British Overseas Airways Corporation , the national airline, were about to cease. This would leave some qualified pilots that were not eligible for service commitment for age reasons, spare. A director of BOAC, Gerrad d'Erlanger, suggested using these pilots as a nucleus for a civil organisation in second line flying duties. These would include ferrying aircraft from manufacturers to operational airfields, medical supply and VIP transport. Being a civilian organisation this would also allow flights to neutral countries without the risk of the aircraft being impounded by the authorities.

On the founding of this organisation on the 11th September 1939, the headquarters were set up at the BOAC flying club at White Waltham airfield. Men and women pilots holding an 'A' grade flying licence and having completed 250 hours flying were eligible to join. Most of their work was to be ferrying aircraft and pilots were graded into classes of aircraft that matched their experience as follows.

> Class I Single engined aircraft
> Class II Fast single engined aircraft
> Class III Light twin engined aircraft
> Class IV Heavy twin engined aircraft
> Class V Four engined aircraft
> Class VI Flying boats

The men and women of the Air Transport Auxiliary made a vital contribution to the war effort, flying in all weathers and were not immune to marauding enemy aircraft. When they were disbanded on the 30th November 1945, it was estimated that 414,984 flying hours had been achieved in 309,011 aircraft.

The uniform chosen for the ATA was cut in dark blue cloth and of similar design to the RAF No 1 uniform with black buttons showing the letters ATA beneath an astral crown. For flying duties, navy blue overalls were worn. Badges reflected the civilian origins and were in gold wire on a black or dark blue melton patch.

A ranking system was used and was shown as cuff rings in various combinations and three thickness. These appeared in gold lace on coloured tape backing, denoting the duties of the wearer. Dark blue for flying, purple for MT, engineering and technical, red for medical, green administration and white operations. Ranks were as follows:

> Commodore 1 thick lace
> Senior Commander 4 medium
> Commander 3 medium
> Captain 1 thin between 2 medium
> Flight Captain 1 thin between 2 medium
> First Officer 2 medium
> Second Officer 1 medium 1 thin
> Third Officer 1 medium
> Cadet 1 thin

Flight Engineers wore three chevrons of black tape on each upper arm.

BADGES
Cap Badge

Three different badge designs for the cap appear to have been worn,

these also being produced in a smaller size for use on the field service cap. The original two badges were embroidered in gold wire, but later a gilt metal version was used. The designs were as follows :

1)Centrally the capital letters ATA within a double rope border, surrounding this twin laurel leaves ensigned by an eagle. All detail on a black,padded, shaped patch.

2)Centrally the capital letters ATA,surrounded by laurel leaves and broken at the top by an eagle. All detail on a black patch.

3)A gilt metal badge showing an oval rope ring encasing an elongated capital letter 'T' flanked by small capital 'A's. Surrounding these,laurel leaves ensigned by an eagle with below a scroll bearing the Latin motto 'Aetheris Avidi'.

Flying Badges

Pilots' Wings :- An oval rope ring containing an elongated letter 'T' flanked by the capital letters 'A', with wings extending on either side. Detail in gold wire or beige silk on a black, shaped melton patch

Flight Engineers' half brevet:
Central design as above,with Half Brevet one wing extending to the

right (as viewed). Detail in grey rayon on a black shaped, melton patch.

Flight Engineers' half brevet II:
As above,with the addition of a twin bladed propeller above.

Senior Flight Engineers' arm badge:
As cap badge No 2, with above a twin bladed propeller. All on a black shaped patch.

Various collar badges were also used. These depicting variations of the cap badge and were either embroidered in gold thread or made of white metal.

FERRY COMMAND

In addition to internal delivery flights from manufacture to Service squadrons, many aircraft were being produced for the United Kingdom in America and Canada. Crating and shipping these aircraft was often impractical, as well as vulnerable to enemy anti-shipping attacks. Aircraft also needed to be supplied to Australia and the Far East. With this in mind Ferry Command was formed in June 1941, to fly these aircraft across the Atlantic and Pacific Oceans. Aircrews, like those of the ATA, were made up of civilian Airline pilots, with no service liability, and women with the appropriate flying experience. In March of 1943, Ferry Command became RAF Transport Command and it's personnel were supplemented by RAF and Allied aircrews.

Uniform information on Ferry Command is scarce, but it would appear to have started as a branch of the ATA and as such, originally the uniforms would have been the same. Later, in March 1943, Ferry Command was renamed RAF Transport Command and it is presumed that RAF uniform would have gradually prevailed.

BADGES
FERRY COMMAND
Pilots' Wings:
version 1
Centrally a circle enclosing a gold thread field, with the RAF monogram. This being surmounted by an astral crown, and beneath the circle, a yellow thread, blank banner. Wings extend from both sides

of the circle. All detail, apart from the banner, in gold wire, on a black woollen shaped patch.

Pilots' Wings:
version 2
Basic layout of detail as above. Wings of softer design and central field in blue. The banner is also blue and bears the words 'TRANSPORT COMMAND'. Detail in gold thread on a black, shaped patch.

Flight Engineers:
Both versions used are half brevet designed as the pilots' wings, only with one wing extending to the right, as viewed,also on a black shaped patch.

TRANSPORT COMMAND
Pilots' Wings:
version 1
Twin twisted rings enclose centrally the RAF monogram on a blue field, with above an astral crown. Below the circle a royal blue banner with the Words 'TRANSPORT COMMAND'. Wings extend
from both sides of the rings. Detail in silver wire on a D/blue, shaped patch, 83mm across.

Pilots' Wings:
version 2
General design as above, a smaller rounder badge, 75mm across, with the central circle in smaller proportions. The banner beneath the circle in this case denoting the squadron number and reading 'ATC 47' etc. Silver wire detail on a D/blue patch.

Flight Engineers:
Both versions as pilots' wings designs above, only with one wing extending to the right, as viewed. Both in silver wire detail on a black, shaped patch.

CHAPTER XI

AIR DEFENCE CADET CORPS

By the end of the First World War in 1918, aviation had grown from a novelty hobby to an important part of His Majesty's fighting services. Aircraft types were being converted from war machines to civil transport, carrying post and passengers. Flying was the advance of the age and as such created great interest in the youth of the day.

As time progressed, two ex-RFC Airmen, Charlie Longman and Bob Weller formed an unofficial club to promote the interest of local boys where they lived in Bournemouth. They made visits to local airfields and aircraft manufacturers and taught the boys how aircraft flew, by the use of models. From a nucleus of six boys in 1928, the now named Bounemouth Young Air League had expanded to forty boys in 1929 and had received recognition from the Air League of the British Empire. The BYAL had by now also created a uniform, which consisted of a double breasted blazer, grey flannel trousers and a dark blue stiff peaked cap. On the pocket of the blazer, a silver winged emblem was attached.

The idea of creating a national youth organisation to promote aviation interest and training percolated slowly though the committees of the Air League until finally in December 1937 Air Commodore J.A. Chamier, the secretary General of the League, proposed the formation of an Air Cadet Corps, for pre service aviation training.

The Air Ministry saw the potential for the recruitment of young men who already had basic skills and assisted in gaining approval for the Corps formation. With a promise of funds from private sources and a capitalisation grant of five shilling per head, all was ready for the formation of an official air cadet corps to be named the 'Air Defence Cadet Corps'.

The first ADCC squadron to be formed was at Leicester in July 1938, closely followed by one at Watford. Although uniform supplies were intended, many Cadets had to buy their own uniforms, at fifteen shillings each, by paying six pence per week.

The uniform was made of blue grey material and worn with a type of Field Service cap, with the ADCC badge attached to the right side (as viewed). The single breasted, five buttoned tunic was worn fastened to the neck and around the waist a belt of the same material was closed by a round, chrome buckle. Breast pockets had buttoned flaps and straps were attached to the shoulders. B/grey trousers, black shoes and socks completed the uniform.

Officers served on Air League Commissions and received no pay or expenses. They wore the RAF uniform jacket of the time, although the trousers had turn ups, with a FSC and ADCC badges. Rank was shown by silver cuff lace.

BADGES

OFFICERS
Cap Badge:
A chrome metal badge, consisting of a pitted circle with raised edges. Within the top the words 'AIR DEFENCE' and in the lower curve 'CADET CORPS' in raised letters. Across the centre a 'speed bird' in blue enamel. Three quarters of the outer circle edge being surrounded by chrome laurel sprigs.

Collar Badge:
A chrome metal 'C'. (Not universally worn).

CADETS
Cap Badge:
The same as the officers above, but without the laurel sprigs.

Squadron Number: A chrome metal number, with 'F' attached if a founder sqn. Worn on the shoulder strap and fixed by two, eyed pins though a backing plate.

The RAF Museum has an example of an embroidered arm badge,but it is not known whether this example was official or privately made.

Ranks: Ranks appear to have followed the Army, from Lance Corporals to Sergeant, there being no Flight Sergeant or Cadet Warrant Officer (Chevrons being L/blue on blue/grey backing).

Lapel Badge:
A miniature of the Cadets cap badge,with a button hole,horse shoe fixing.

Belt buckle:
A two piece chrome buckle, with the left belt loop attached to a slotted circular plate. The right side of the buckle was domed and circular to slot through the left plate and had an embossed 'speed- bird' on a pitted background.

Buttons: These are chrome and show an embossed 'speedbird' with the raised letters 'A.D.C.C.' above.

Civilian Brassard: As worn by the Civil Air Guard.

AIR TRAINING CORPS

Success of the ADCC, as it progressed through the difficult early War years and the close bond that formed between the RAF and the ADCC, as the cadets visited and assisted on Royal Air Force Stations, prompted the Air Ministry to look at the Corps to be integrated into the Royal Air Force,as part of the Training Command. Already cadets entering the Royal Air Force and the Fleet Air Arm had shown the authorities that early training of skills in aviation, mechanics and Service life etc, quickened the recruits' entry into active service.

The first step was staff and following a press and radio broadcast by the Secretary for Air, on the 10th January 1941, for potential officers and staff to contact the Headquarters of this new formation, at Stanmore, hundreds of copies of the new regulations on running the newly formed Squadrons (Air Ministry Publication 1919), were dispatched to Local Authorities and cities throughout the Country. This publication, although modified, is still the base document in use today.

The Air Training Corps was officially established on the 5th February 1941, with King George VI agreeing to become the Corps Commander in Chief and issuing a Royal Warrant for the Corps formation. Within the first month, 400 Squadrons had been formed, including nearly 200 ADCC squadrons which were integrated into the ATC.

As the War progressed, Air Crew requirements slowed down and the main purpose of the ATC slowly decreased. There was a concern that the Air Training Corps may be disbanded. These fears were overcome in 1945 with the announcement that the Air Training Corps would become part of Reserve Command and fully integrated into the Royal Air Force.

With the return of peace, a new aim was needed for the Corps and in 1947 a new Royal Warrant was issued that read as follows:-

1) To promote and encourage among young men a practical interest in aviation and to fit them to serve their Country in Our Air Force, its reserves and auxiliaries, and also in the Air Branch of Our Navy or in Our Army.

2) To provide training which will be useful in Our Service and Civil life.

3) To foster the spirit of adventure, to promote sport and pastimes in healthy rivalry and to develop the qualities of mind and body which go to the making of a leader and a good citizen.

Although this Warrant was slightly modified in 1967, the basic aims continue today. The cadets working though a syllabus of training and progressing though exams to gain proficiency badges and to become cadet NCO's.

Air Training Corps adult staff are all volunteers and except for occasional activities, unpaid. They comprise uniformed Officers and Warrant officers as well as civilian instructors. The Officers today are commissioned in the Volunteer Reserve (Training) branch of the Royal Air Force, but have no war commitment. Air Ministry Order administrative 567 of 24th July 1941, authorised Air Training Corps Officers to wear Royal Air Force officers' uniform, as in AP 1388 Dress Regulations, and rank braid, although without the SD cap, as a Field

Service Cap was to be worn. Warrant Officers, or Adult Warrant Officers as they are officially termed, have worn the Non-Commissioned RAF uniform of the time. They hold a unique position in the Corps as they are 'uniformed civilians', and as such are not part of the Reserve Forces, although come under Dress and Discipline Regulations, with many holding officer posts at squadron level. Civilian Instructors, as the name suggests, are a dedicated band of helpers, who instruct cadets in various subjects of which they have skills. They do not wear uniform, although originally they were issued with fawn dust coats.

CADETS

There have been two major changes in cadet uniform since the Corps formation. These have followed uniform changes of their parent Service and tend to follow once the Royal Air Force have re-kitted. This often meant that cadets of a Squadron would be found in different uniforms at the same time.

The uniform chosen for the new cadet force of 1941 closely followed it's predecessor. The five buttoned, single breasted tunic, in B/G , retained the buttoned breast pockets and was closed at the neck by a hook and eye. The belt was replaced by an attached half belt, with a two pronged gilt metal buckle. Buttons remained chrome and showed a raised falcon in flight, with above the letters A.T.C.. Blue grey trousers were worn and the Field Service cap was retained. Shoes were black. Demand outstripped supply at first, cadets being issued with dark blue boiler suits. In November 1946, B/G serge greatcoats were added to the cadets equipment scale. These were as the RAF Airmen's issue with large ATC chrome buttons.

During 1953, the now familiar beret replaced the field service cap for cadets, as a prelude to a major change in uniform which followed in December 1958. This was the Battle Dress (War Service Dress) and trousers in blue grey woollen material. Very itchy in hot weather! This uniform did however allow for a shirt and tie to be worn under the open neck. The cadets were issued with two detachable collars and one shirt.

With the arrival of the 70's a more casual and comfortable approach was being introduced into the Services and a crew necked woollen, jersey with elbow and shoulder patches was introduced, which also reduced the cost of the uniform. These uniforms filtered down to the cadets in 1977 and apart from a change to a V neck, remain today, the badges being worn on a B/G brassard worn on the upper right sleeve of the jersey.

BADGES

Heraldic Crest

The Heraldic Crest created for the Air Training Corps, depicts a falcon with raised wings, on a silver field within a red circlet. The circlet bearing the words 'AIR TRAINING CORPS' in gold lettering, at the bottom a star, also in gold. Above the circlet a gold astral crown. The Corps' motto 'Venture Aventure'is inscribed on a silver scroll beneath the circlet.

A crownless version in gilt metal appears in four of the Corps badges.

OFFICERS

Headgear badges and rank lace are as for Royal Air Force officers, with the addion of the following:

Tunic

Lapel Badge:
(1941-1944)
3/8 inch high, brass 'VR' letters.

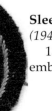

Sleeve Badge:
(1941-1944)
13/4inch circular,black patch with within an embroidered ring the letters A.T.C., all in L/blue.

Greatcoat
Shoulder Badge:
(1941-1944)
   1/2 inch high brass 'VR' letters.

Tunic Collar Badge:
(1944-1947)
   A gilt metal, crownless Heraldic Crest.

Tunic Collar Badge:
(1947-present)
(AMO A 754)
   Brass or anodised aluminium letters 'VR' appear-
ing on the top bar of a 'T'(Volunteer Reserve
Training),12mm high.

Greatcoat.:
   As above, 24 mm high. (obsolete)

ADULT WARRANT OFFICER

Rank Badge:
(1941-1944)
   A laurel wreath, broken at the top, with
across the centre a flying eagle all embroidered in L/blue, on a round 2 inch patch.
Issued in pairs, the eagle'flying to the rear.

Rank Badge:
(1958-present)
AMO A 356 31 Dec 1958
   Q/Crown 48mm high embroidered
in L/blue, on a blue grey or black
shaped patch.

Shoulder Slide:
   40mm Q/C embroidered in silver on
a D/blue 70mm slide.

Until July 1988, Adult Warrant Officers, having served eight years and passed the appropriate courses, were allowed to 're-badge' and wear the Royal Air Force Warrant Officers' Royal Coat of Arms badge. Unannounced, and thereby depriving qualifying Adult Warrant Officers from attending the final course before the date, course 4/88 of the Air Cadet Training Centre, at RAF Newton, became the last course to re-badge. Only Adult Warrant Officers passing the criteria before this date or Ex-servicemen/women with the equivalent rank are now permitted to wear the Royal Coat of Arms rank badge.

From 1958, all Adult Warrant Officers wear gilded metal, anodised ATC letters beneath the rank shoulder slide or the lapels of the No 1 jacket. Headgear badges are as for Royal Air Force Warrant Officers.

CIVILIAN INSTRUCTORS

Chaplains

ATC squadrons encourage local civilian clergymen to become the squadron chaplain and a stole badge is available for them to wear. This shows a chrome cross patt'ee with a grooved appearance and plain border and center. Superimposed on the middle is a crownless version of the heraldic crest in gilt metal.

Civilian Instructors

Civilian Instructors have had two forms of identification available for them to wear, although they are seldom worn.

Arm Band:

A 90mm cloth band in mid blue, with a white border. Printed in th upper border the words 'AIR TRAINING CORPS' and in the lower border 'INSTRUCTOR'. The centre shows a printed version of the cadets beret badge.

Lapel Badge:

A crownless version of the heraldic crest, with between the raised wings the letters 'CI'. All in gilt metal.

Welfare Lapel Badge

This badge is presented to civilian personnel who have been particularly helpful to the Corps. It is a crownless version of the heraldic badge in gilt metal, with the initial 'W' between the wings.

CADETS

Beret & FSC Badge:
A 38mm diameter chrome badge of a circlet containing the words 'AIR TRAINING CORPS' on a pitted background. Across the central void a flying falcon. (attached by split pin and eyes.)

Field Service Cap Badge:
(War economy)
As above in silver grey plastic with a two prong metal fixing.

Squadron shoulder title:
(1941-1963)
An oval B/G patch, with a L/blue border. Across the top 'AIR TRAINING CORPS'and along the botton 'SQUADRON' in L/blue thread. Centrally squadron number and location

Squadron shoulder title:
(War economy)
As above, printed in white on a black oval patch.

Squadron shoulder title:
(1963-1977)
A curved,135mm, B/G or D/blue patch with'AIR TRAINING CORPS' in L/blue. A separate oblong B/grey or D/blue patch worn beneath the title with the Squadron number in L/blue thread.

Squadron shoulder title:
(1977-present)
A curved,D/blue,75mm patch with 'AIR TRAINING'across the top and 'CORPS'beneath

Civilian dress lapel:
A miniature of the cadets badge in chrome.Early badges have a horse-shoe button hole fixing. The latter a pin and hook. A war economy badge was made of silver grey plastic.

Proficiency Badges

First Class Cadet:
(since 1941)
A four pointed star in L/blue silk on a B/G or D/blue shaped patch.

Leading Cadet:
(since 1942)
A four bladed propeller, embroidered in L/blue on a D/blue or B/G patch.

Senior Cadet:
A four bladed propeller, with a four pointed star superimposed on the centre. On a B/G or D/blue square patch.

Staff Cadet:
(1968-present)
A yellow lanyard. (Worn on the left shoulder.)

Rank Badges

Cadets of today's Air Training Corps use the same rank structure as the Royal Air Force and the same badges, with the exception of Cadet Warrant Officer, Flight Sergeant, Sergeant, Corporal chevrons were originally as used by the RAF Apprentices. On today's jersey shoulder slides are worn. Until 1958 cadets were not permitted to wear the Royal crown for Flight Sergeants. Their place was taken by an eagle emblem.

Flight Sergeant:
(1941-1958)
A 1 1/2 inch circular patch in B/G cloth with a L/blue embroidered border. Across the centre a flying eagle also embroidered in L/blue thread. (Issued in pairs, eagle 'flying' to the rear).

Cadet Warrant Officer:
(since 1958)
A B/grey or D/blue cauldron shaped patch with a Q/C surrounded by laurel sprigs in L/blue embroidery.

Cadet Warrant Officer shoulder slides:
Detail as above, silver embroidery on a D/blue 65mm slide.

FLYING

As the Air Training Corps was originally established to speed recruits into the Royal Air Force as aircrew, flying training has always been at the forefront of ATC activities. Air experience flights, in RAF aircraft, have always taken place to give the cadets a feel for flying. These are often as passengers, although flights in light aircraft allowed hands on training. In 1958 the cadets were presented with a dedicated fleet of Chipmunk aircraft for air experience flying, these have now been replaced by the Bulldog.

Selected cadets are able to participate in a Flying Scholarship course to achieve a Private Pilots' Licence (PPL) or an Air Cadet Pilot Navigation scheme. Those successfully passing these courses are presented with badges, worn on the left jersey shoulder patch, as follows.

Pilots' Wings:

Centrally the letters 'FS' within a D/blue ring, with above 'AIR' and below 'CADETS' in yellow. The central ring supported by outstretched wings. The 'FS' and wings in silver blue. All detail embroidered on a black, 95mm patch This basic design has also appeared in L/blue on a shaped B/G patch.

An earlier badge of the 60's depicted a circle in which the words 'FLYING SCHOLARSHIP' were printed, with the circle supported by raised, outstretched wings.

Navigators' Half wing:

A 95mm, black cloth patch with an off-set D/blue circle enclosing the letter 'N'. Above the circle the word 'AIR' and below 'CADETS' in white thread. Extending from the right of the circle,as viewed, a stylised wing. Wing and 'N' in silver blue embroidery.

A badge is also available for Staff Cadets who assist on the flying courses. Worn on the flying coveralls, it shows the words 'FLIGHT STAFF CADET' in silver blue thread on a 95mm black patch.

GLIDING

Another way to introduce cadets into flying is gliding and in 1943 the Air Training Corps received it's own small fleet of gliders, with an establishment of dedicated gliding schools. By 1945, 84 gliding schools had been established with a strength of 200 craft. Today, as with cadet strength, fewer gliding schools remain, with 27 Volunteer Gliding Schools (VGS) throughout the country being administered from RAF Cranwell.

Cadets achieving passes at various levels of training, wear a winged badge on the left shoulder patch of their uniform jersey. Over the years this badge has varied in quality and colour, although following the same basic design and are as follows.

Gliding Wings

Cadets
Course Completion:
 A blue embroidered, angular flying bird with above 'AIR' and beneath 'CADETS', in white thread on a D/blue ,shaped patch.

From July 1998:
 Blue embroidered wings, with centrally a ring enclosing the letters'GS'in white thread above the ring 'AIR' and below 'CADETS' also in white thread. On a 60mm X 25mm D/blue shaped patch.

Solo Completed:
 As angular bird, embroidered in silver/blue rayon thread. (Earlier badge, cotton on melton patch).

From July 1998:
 Silver embroidered wings, with a white 'S' centrally.

Advanced Pilot:
 As angular,bird embroidered in yellow rayon thread.

From July 1998:
 Gold embroidered wings, with a white 'A' centrally.

Adult Staff
Gliding Instructor:
   Centrally a letter 'G' within a ring,supported by out-stretched wings and surmounted by a Q/C. All detail embroidered in L/blue on a D/blue shaped patch.

Glider Pilot:
   As above without crown.

RIFLE SHOOTING

Competition rifle shooting instils discipline and pride and cadets have been trained and coached to obtain high scores at inter-service competitions for many years. Initial training starts with air rifles and progresses through miniature .22 calibre to a full bore service discipline.

Badges are awarded for different grouping, scoring and competition achievements and requalified for annually.

ATC Marksman: A Lee Enfield rifle pointing to the left, as viewed, with above a Q/Crown with red cushion. L/blue on a D/blue or blue grey shaped patch.

RAF Marksman: Crossed Enfield rifles embroidered in light blue on a shaped, B/G or D/blue patch.

Cadet One Hundred: A circular D/blue patch with border, showing an archer and rifleman. Above the figures 'NRA CADET HUNDRED' and below 'SIT PERPETUAM', with the year awarded. Embroidered in white

Awarded by the National Rifle Association for the best one hundred scores at the annual Bisley competition.

BANDS

Bands are encouraged at Squadron level and in some areas are formed into larger Wing Bands. Many smaller bands comprise drum and bugles with lyres, tom toms etc supplementing them. In July 1996, the first National Air Training Corps Band was formed, for one week, from cadets from all over the British Isles. These cadets played in concert with the Royal Air Force Regiment Band at RAF Northolt, following rehearsals at the camp. The outstanding success and musical skill achieved during this time has now resulted in cadets playing annually with a Royal Air Force Band.

Badges are in chrome and depict a side drum, crossed bugles, bagpipes and a lyre resting on oak leaves. All are fixed to the sleeve by pin and eyes.

Communicator Badge

Most ATC badges have been with the Corps for many years and it is refreshing to see a new qualification badge appear. Promulgated in Corps Wing Routine Orders of the 9th March 1997, the new Communicator badge was designed by Malcolm Wood, a Wing Radio Officer, and modified by the author. The Malcolm badge?. It requires a high standard of proficiency in the operation and skills of Radio Communication equipment and it is hoped that it will soon grace the uniforms of many cadets.

Communicator Badge:
Four L/blue lightning bolts emitting from a central bright blue letter 'C'. Embroidered on a 35mm wide D/blue, oblong patch.

Deferred Service

A War-time badge worn beneath the squadron number, which denoted that the wearer would enter the RAF on reaching call up age. It showed the letter 'V' in L/blue on a black background. A Naval Youth Entry badge was also worn.

COMBINED CADET FORCE

As early as the 1860's Public Schools have formed training units for pupils as a preliminary start to an Army life. As time progressed, an Officer Training Corps was formed by 1908 to train potential officers for the Army and Navy. During World War One many Grammar Schools formed Army Cadet Force units (ACF) and these were expanded during 1940 to include cadets for the Royal Navy and the Royal Air Force.

Following the War, a rationalisation of these Cadet Forces was sought and schools that had pupils of 17 years and over were invited to join the Combined Cadet Force, a newly formed organisation which combined all school Cadet units together for leadership training for the three Services.

The Air section cadets enjoy most of the activities of the Air Training Corps cadets, such as flying, rifle shooting, sports etc and are taught related subjects such as airframes, engines, electronics etc, although they still keep a strong bias on leadership qualities.

As with the Air Training Corps the Officers are members of the Royal Air Force Volunteer Reserve (Training) and wear the same uniform and badges.

The cadets wear the same uniform, rank and proficiency badges as the ATC cadets. They do however wear the RAF anodised cap badge on their berets and a dedicated squadron identification badge which shows on a red patch the school name and C.C.F. in light blue lettering.

Chapter XII

THE ROYAL OBSERVER CORPS

As Hauptmann Linnarz glided over the English coast line on the night of the 31st May 1915, he was unaware that his actions that night would cause the creation of the world's first early warning system. No searchlights or interceptor fighters had located the Army airship LZ 38 he commanded as he flew above Essex towards East London. Here he dropped his bombs, which resulted in the death of seven people and an estimated £18,000 of damage.

Although this was not the first bombing of British soil, raids had been made on coastal and Naval installations, the enemy had reached the capital of England! and something had to be done. At the time, any enemy airships that were spotted by Army units, Police or even railway stations, (airships could us railway lines for navigation), were reported to Horse Guards in London. Sometimes even by post. This meant that although enemy airship raids were known, none were spotted or reported in time to be shot down by aircraft, or the twelve anti-aircraft guns protecting London. The only warnings given to the public within a sixty mile radius of London were localised and given by police cycling the streets, blowing whistles.

In response to this raid more aircraft and guns were allocated to the London area and more importantly, 200 observer positions were established, to try to locate the airships and give adequate warning for the defences and public. These measures were not to prove successful and in the early part of 1917 a greater threat appeared in the form of the German twin engined, Gotha bomber. On the 13th June 1917, twelve of these aircraft bombed London and escaped without interception.

To improve the situation the War Office engaged Lieutenant General Smuts, the instigator of the RAF, to prepare a study document to improve the state of the defences. Apart from more anti-aircraft guns and searchlights a better spotting and reporting system was required. Smuts concluded that the London defences needed to be co-ordinated by one senior officer and recalled General E.B. Ashmore from Ypres, where he commanded an artillery division, and placed him in command of the London Air Defence Area (LADA), which came into being on the 31st July 1917.

A network of observer posts was set up, reporting to a central control, where the telephone reports were co-ordinated.

As communications improved and plotting tables developed, more success was achieved in the interception and turning back of air raids. More fighter squadrons were also made available and balloon aprons were tried. These measures were tested to the full with the arrival in December of the four engined Staaken German bomber.

It was during this time, that the plotting table, markers and coloured sectioned clock developed what with a few modifications became the standard control tools for Fighter Command's control plotting during the Second World War.

By the 29th October 1925, the official inauguration of the Observer Corps, the observer organisation had been resurrected, under the utmost secrecy, to man a network of observation posts in Kent, Hamshire and Essex. Telephone lines were laid to these positions and connected to a central control, usually in a GPO telephone exchange. The observer posts would report passing aircraft, seen or heard and with rudimentary instruments would access direction and height. Central control would take this information and by triangulation estimate the aircraft's flight path.

As the Observer Corps expanded to the north and west, to eventually become 18 groups covering the Country, the Government at the time thought that the Air Ministry would be better suited to supervise this expansion. A transfer from War Office control was agreed, to take place on the 1st January 1929.

During 1929 the new commandant, Air Commodore G.A.D. Masterman, consolidated the Corps and created additional posts integrated into Naval and Coast Guard stations. It was during a conference held in October,that the first of the Observer Corps badges was announced. This being a lapel badge. It should be remembered that apart from the Special Constables uniform, Corps members wore civilian clothes on duty, with the addion of an arm brassard. This badge could be purchased by members, but latter qualified observers were allowed free issue.

It was during this time that the Army had begun experimenting with the idea of locating aircraft by sound location, to warn their anti-aircraft and search light batteries. Huge 'sound mirrors' made of concrete, had been erected on marches near the Kent coast, some measuring 26ft (8m) high by 200ft (61m) wide. Although many tests were carried out they were moderately successful, being able to improve detection of about three times that of the human ear and with no height information. By the end of 1936, detection by this method had been stopped as it had become to complicated for active field use.

In 1935, the Radio Research station at Slough, had demonstrated that with the use of radio waves it was possible to detect and locate aircraft and the development of Radio Direction Finding (RADAR) began.

By August 1938, five of these radar stations had been positioned around the south east coast of England. These were called Chain Home stations and transmitted and received radio waves through large aerial arrays suspended from steel towers. An overall 'floodlight' pattern was received on a cathode ray tube up to a range of 200 miles at 35,000 feet. Skilled operators were required to distinguish incoming aircraft from other screen interference. It was found that below 3,000 feet this system was not effective and 'Chain Home Low' was developed, using a 'beam' type radar that could pick up aircraft at 1,000 ft from a distance of 180 miles.

Although these systems became essential in warning fighter and other defence units in the Battle of Britain, radar was still in it's infancy. Human observers were still needed to spot attacks flying below 1,000 feet and raids flying over the mainland, behind the radar screen.

It was feared during these technological developments that the Observer Corps could become obsolete, but with radar gaps and the prominent fear the Country had at the time of a parachute invasion, the Corps moved towards the impending war with renewed vigour. During the early days of the Second World War, the expected bombing raids did not arrive and the Observer Corps posts found themselves assisting British pilots lost in bad weather, by reporting their presents to RAF control, who were able to talk them down.

Following the evacuation of the British Expeditionary Force and their allies from the beaches of Dunkirk, in May and June of 1940, the threat of invasion became a reality. Many of the Observer Corps posts were in prominent positions around the English coast and as such would be in the front line. Rudimentary defences were built around some posts and apart from the police, the Observer Corps became the first civilian group to be issued with rifles and ammunition, two rifles per post. It was at this time that AMWO A 423 authorised RAF blue overalls to be worn with a dark blue beret and new corps badge. Officers' were to wear Royal Air Force officers uniforms with ROC cap and collar badges. All the ROC staff were issued with Observer Corps arm bands and identity cards.

The Battle of Britain saw the Corps posts as information outposts, for as well as reporting attacking enemy aircraft ,they reported the location of baled out aircrews, friend or foe, crashed aircraft and bomb damage. As Luftwaffe attempts to destroy the Royal Air Force failed, the air battle progressed to the bombing of cities, the Blitz. The Observer Corps continued to track the incoming bombing fleets, even when the enemy changed course inland, to confuse the defences.

In January 1941 an Observer Corps committee sat with members of Government to examine the service that the Corps had given over the past sixteen months of war, to improve it's civilian 'part time' status and role for the future. Various new titles were suggested for the Corps and applications were made to the King for the grant of a Royal prefix. Inquiries were made to other civilian organisations, in case of seniority objections, but none were received and so on the 9th April 1941 parliament announced that the prefix 'Royal' had been granted by the King. In addion, a one piece uniform, that would go over civilian clothing was to be issued. This coverall was never popular, being a very ample fit. A large breast badge was also promulgated, showing a RAF eagle surmounted by a crown. Fortunately in early 1942, AMWO A 64 authorised Corps members to wear a Royal Air Force blue, Heavy Duty Dress,(Battle Dress) with ROC badges and buttons, to replace this baggy style.

As the war years progressed, manpower of the Corps was dissipated. By 1941 the young men had been called up for active service and although men between 35-40 had their call-up deferred, the Armed Services needed every man. This left a severe shortage in the Corps and in July 1941 women were recruited into the ROC. At first their work was in the central control centres, but gradually they assisted at spotting posts. These posts had always been primitive, being at best a garden shed, and it is a credit to these ladies that they managed in these conditions.

The recruitment of women also had it's down side. When a new Corps Commander was appointed in June 1942, one of his recommendations was that men over fifty were required to leave the Service. Previously, being a civilian volunteer service there was no upper age limit.

Now that the Battle Dress uniforms were being issued and a closer liaison with the Royal Air Force was becoming essential, the Royal Observer Corps took on a more military look. A rank structure, with new badges, was created which was used until its final demise. Officers were to wear RAF officers uniform with ranks worn in the same manner, only in a midnight-blue colour. Two 'NCO' ranks were created, that of Chief Observer and Leading Observer. More training and tests were given, for plotting and aircraft recognition, which had three standards, Basic, Intermediate and Master. The latter being awarded a Spitfire proficiency arm badge.

By 1943, the tests had become compulsory for control centre and post staff. These tests now included oral and written tests on ROC procedure, current orders, as well as aircraft recognition and descriptions of 50 aircraft. To obtain the Master 'Spitfire' badge required the Observer to recognise correctly 180 views of aircraft out of a possible 200, with mark numbers. A very high standard. From this date annual tests were taken.

In early 1944 the Country was building up strength for Operation 'Overlord', the invasion of occupied France. Spasmodic bombing raids were occurring on cities and the coastal ports and the ROC was keeping up with the recognition of new German aircraft.

The Authorities responsible for planning the invasion, among other things, were concerned by the poor aircraft recognition standards aboard many of the defensively equipped merchant ships (DEMS). During the invasion of Sicily in July 1943, American ships had opened fire on Allied aircraft and many ships regarded any aircraft as hostile. To remedy this situation the Admiralty requested the assistance of 630 ROC observers. Of the 1,376 who volunteered, 796 were eventually selected, to serve in pairs onboard landing ships, transports and American vessels. The Observers chosen were to be given the rank of Petty Officer for one month's duration with two additional months if required and were to receive pay of £1 per day. They would wear their ROC uniforms with the addion of a navy blue brassard showing the initials 'RN' and a 'Seaborne' shoulder badge. These Observers served with distinction during D-day and after and their aircraft recognition skills were responsible for saving many a RAF pilot's demise from the guns of friendly forces.

The land based Observers were now plotting the flight paths of a new threat, the sleek lines of the V1 flying bomb. The first flight being spotted over Dymchurch, Kent on the night of 13th June 1944. These attacks being a prologue to the V2 rocket, against which there was no defence and which only stopped when their launch sites were overrun.

With the Allied forces advancing across Europe, enemy aircraft became scarce over England and the Royal Observer Corps posts ended the war years assisting lost and damaged Allied aircraft find their airfields.

Following the end of hostilities in Europe, the Royal Observer Corps stood down on the 12th May 1945. It's personnel returned to civilian life, leaving only the Headquarters at Stanmore and area offices to administer the centres. Corps members still contrived to hold their own local meetings and various official visits were organised to air shows. By 1946, the Committee for Post-War Organisation had decided that there was a place for the ROC in a peace time role. Full-time officers were given a three year contract and war time Observers were invited to re-join. On the 1st January 1947 the Corps was reinstated, with an established strength laid down as 57 full-time officers, 415 part-time officers and 27,901 part-time Observers. This strength commitment could not be filled by war time Observers and recruitment and training was started.

Many of the post sites had either been vandalised or had fallen into disrepair and it was found necessary to commence a programme of standardising the construction of post sites, these becoming a raised wooden shed structure with an open observation platform at one end. In the interim, 'Sidcote' flying suits, greatcoats and ground sheet capes were issued to the hapless post observers.

The Soviet Union was by now showing an aggressive stance towards the West and following the closure of the border around Berlin in June 1948, the Berlin airlift continued until October 1949. New Soviet aircraft types were also becoming a serious threat to the security of the Western Powers. These ranged from B29 copies to the new Yak and Mig fighters. Aircraft recognition was once again becoming a priority, as UK defences prepared for possible attack. At the time it was found that the radar equipment available was not adequate to detect low flying, high speed aircraft and the ability of trained Observers to accurately recognise and report an aircraft quickly was essential. Aircraft recognition was intensified.

These measures soon became abruptly inadequate with the news in

September 1949 that the Soviet Union had successfully exploded a nucular device. This combined with the ever expanding Soviet bomber force increased the threat to the Western Powers, who had grouped together on the 4th April 1949 to form the North Atlantic Treaty Organisation (NATO).

In the UK, the ROC were soon to take on additional reporting tasks. By 1954, concern was mounting over the result of a Nuclear attack on Great Britain. Apart from the destruction at the centre of a nuclear explosion, the fall-out would drift with the prevailing wind and contaminate the ground beneath. It was necessary to plot this nuclear cloud and as the ROC had a national coverage of posts, as well as an inbuilt reporting system, the Home Office requested their help.

To safely report any fall-out pattern, its was necessary to install instuments to monitor bomb-burst strength, radio active fall-out and wind direction. This information had then to be transmitted to control centres. For the personnel manning these positions, protection had to be provided. In 1956 constuction of underground posts started, which eventually was to lead to a total of 1,563 being built, as well as 31 control centres. These posts were made of reinforced concrete sunk underground. Beneath a 7.5 inch slab, accommodation for four observers was housed in a 15ft x 7ft room. This area contained two bunk beds, canvas chairs, a cupboard and table, as well as monitoring instuments. A chemical toilet was provide in a separate room at one end of the post, adjacent to the 15ft entrance shaft. Air was provided by two filtered, louvered ventilators. Post and centres were connected by telephone, with master area posts holding additional radio equipment.

Since the early 1950's, the ROC had been taken over by Fighter Command and it's members had assisted in manning fighter control centres. By 1965 the Royal Observer Corps had been integrated into the United Kingdom Warning and Monitoring Organisation (UKWMO). The Corps had by now also lost it's principal role of aircraft reporting and was now solely involved in the nuclear reporting task.

Throughout the 70's and 80's advances made in radar, communications and monitoring equipment had meant that posts were becoming less reliant on the human factor. Sensors could automatically sense any change in local conditions and send this information to centres, which were now connected by automatic switching exchanges to any national number. Personnel numbers had been run down and during 1990, the Home Office had re-shaped the emergency defence system though local authorities. On the 10th July 1991, the House of Commons was informed by the Government that the Royal Observer Corps would stand down. Training ceased and over the next 41 months the Corps was run down, to finally disband on the 31st December 1995, with the permanent H/Q staff leaving RAF High Wycombe on the 31st march 1996.

BADGES

When the Observer Corps first became official on the 29th October 1925, it will be remembered that the members were civilians and wore civilian clothes, being only identified as members of the Corps by a Special Constables arm band. Uniforms as such, were not issued to the Corps until 1940, these being coveralls and berets. In 1942 the now new Royal Observer Corps was issued with their first complete uniform, the blue grey, Heavy Duty Dress, which was to be worn until the 70's when the RAF 1972 pattern jacket and trousers were supplied.

Badges of the ROC reflect the Corps' seventy year history and transition from civilian clothing to full uniform.

Heraldic Badge

The Royal Observer Corps heraldic badge depicts an Elizabethan coast watcher in breast plate and helmet holding in his raised, right hand a flaming torch, his left hand shielding his eyes. Laurel sprigs surround the circular centre, which is surmounted by a K/Q crown. Beneath the figure a scroll bears the motto ' FOREWARNED IS FOREARMED'.

Lapel Badge:
(Oct 1925 - 1941)
A gilt metal, circular badge. Centrally a scene in relief, depicting an Elizabethan coast watcher on a cliff, beside a flaming beacon. The scene is bordered in pale blue enamel, with the motto 'FOREWARNED IS

FOREARMED', above and below 'OBSERVER CORPS'.
Lapel Badge:
(April 1941- 1953)
As above in white metal, surmounted by a King's crown, with red cushion.

Lapel Badge:
(1953-1995)
As above with Queen's crown. (Two sizes made)

Arm Band:
This arm band was two inches wide and showed alternating, vertical 1/4 inch blue and 3/4 inch white stripes, with 'OBSERVER CORPS'in 1/2 inch red letters across the centre.

OFFICERS

Officers wore the Royal Air Force No1 uniform, with the buttons changed for ROC ones, showing a raised coast watcher design on a domed, gilt metal button.

Cap Badge:
JAN 1042
AMO A 64
A gilt metal Elizabethan coast watcher and base with below a scroll bearing the motto 'FOREWARNED IS FOREARMED'. This being attached centrally to an oval, padded black parth.The figure has on either side a gold embroidered laurel sprig with above a K/Q crown in gold and red thread.

Head Observer: post:
An oblong, black label with 'HEAD OBSERVER' in L/blue embroidery. (Worn above left breast pocket).

Duty Controller:
(Control centres)
An oblong, black label with 'DUTY CONTROLLER' in L/blue embroidery. (Worn as above).

Ranks *(AMO A 207 March 1943)*

Chief Observer:
Centrally, 3 horizontal bars cupped by laurel sprigs. Embroidered in L/blue on a B/G or black patch.

Chief Observer:
shoulder slide
Detail as above in silver embroidery on a D/blue 50X50 D/blue slide.

Leading Observer:
As chief Observer with 2 horizontal bars.

Leading Observer:
shoulder slide
As above, embroidered in silver thread on a D/blue slide.

Shoulder Titles:
Aug 1950
A curved patch in B/G or black, shaped as the shoulder seam, with 'ROYAL OBSERVER CORPS' embroidered in L/blue.

Group Number:
L/blue embroidered number on a B/G square patch. (Worn beneath the shoulder title).

Collar Badge:
The letters 'ROC' in gilt metal 3/8 inch high, 1/2 inch for greatcoat shoulder straps.

RANK
Officers originally wore a 1 1/2 inch blue braid on their RAF uniform sleeve and greatcoat shoulder strap (AMO A 423, June 1940). In March 1943, AMO A 207, promulgated a rank structure which was similar to the Royal Air Force in position and combination, only in midnight blue lace.

Observer Commodore:
One lace 2 inches wide lace.

Observer Captain:
Four rows of 9/16 inch lace.

Observer Commander:
Three rows of 9/16 inch lace.

Observer Lieutenant Commander:
One 1/4 inch lace between two rows of 9/16 inch lace

Observer Lieutenant: Two rows of 9/16 inch lace

Observer Officer: One row of 9/16 inch lace.

OBSERVERS

Beret Badge:
June 1940
(AMO A 423)
In white metal (later anodised aluminum), depicting the Elizabethan coast watcher surrounded by a laurel wreath, which is pieced at the top by a K/Q crown. The bottom of the wreath is superimposed by a scroll bearing the motto 'FOREWARNED IS FOREARMED'.

HDD Breast Badge:
Jan 1942-Aug 1950
(AMO A 641)
A RAF eagle with above 'ROYAL' and and below 'OBSERVER CORPS' all within a circle surmounted by a King's crown. In L/blue embroidery on a pear shaped black patch.

In December 1943, aircraft recognition and operational regulations tests became compulsory. The annual aircraft recognition test was set and a badge showing the underside of a Spitfire aircraft in L/blue embroidery on a B/G patch, was awarded to those who successfully passed. This badge being worn on the upper, left sleeve. In March 1954 repeated passes by Observers resulted in a system of Spitfire and star badges, to denote the number of passes obtained.

Intermediate Pass: L/blue star on a B/G patch.
1 Annual Pass: L/blue Spitfire badge.
5 Annual Passes: Red Spitfire badge.
10 Annual Passes: Red Spitfire badge above a red star.
15 Annual Passes: Red Spitfire badge above 2 red stars.
20 Annual Passes: Red Spitfire badge above 3 red stars.
25 Annual Passes: Yellow Spitfire badge.

Seaborne Badge:
A curved B/G patch bearing the letters 'SEABOURNE' in 'in L/blue thread. Worn on a D/blue brassard with 'RN' in red.

Buttons worn on the B/G battle dress uniform were as officers only in white metal.

Suggested Reading

Cormack, Andrew, *Royal Air Force 1939-1945*. Ospey

Congdom, Philip, *Per Ardue ad Astra*. Airlift

Herring, Squadron Leader, *Customs and Traditions of the Royal Air Force*. Gale & Polden

Rosignoli, Guido, *Airforce Badges and Insignia of WWII*. Blanford

Carmar, May, *Badges and Insignia of the British Armed Services*. Tanner

Gander, Terry, *Encyclopedia of the Modern Royal Airforce*. Patrick Stephens

Philpott, Bryan, *Challenge in the Air*. Model and Allied Publishing

Wood, Dereck, *Attack Warning Red*. Carmichael & Sweet

Military Modelling Magazine. September/October/November 1985

Acknowledgments

This book could not have been completed without the help I received from the staff of the Royal Air Force Museum, Hendon. I would particularly like to thank Andrew Cormack for his encouragement and help in allowing me to use his reference material.

While researching for this book I was grateful for the assistance afforded to me by members of the Royal Air Force and ex-service men, and I would like to mention Flight Sergeant Gil Singlton for his assistance with band and other current Service badges and Cris Wren, curator of the Battle of Britain control room museum.

A further thank you must be extended to Mrs Blackmore of the Newark Air Museum for the valuable assistance given to me on the Royal Observer Corps. Their collection of ROC badges is the best I have seen.

Malcolm Hobart